In this second edition of his classic work *Foundations of Mathematical Genetics* A.W.F. Edwards gives a definitive account of the basic models of deterministic population genetics together with their historical background. Existing texts in mathematical genetics reveal a continuing need for a careful study of the foundations which this book satisfies, treating the simple deterministic models for random-mating diploid populations in depth without sacrificing clarity of expression. In the new edition coverage has been extended with the provision of a new chapter on the Fundamental Theorem of Natural Selection.

The book is intended for those interested in the mathematical aspects of genetics, ecology and biology. Students of mathematical biology and historians of the subject will find it the definitive statement of the origins of modern mathematical population genetics.

A. W. F. EDWARDS is Reader in Biometry in the University of Cambridge. His other publications include *Likelihood* (1972, 1992) and *Pascal's Arithmetical Triangle* (1987).

W0042205

FOUNDATIONS OF MATHEMATICAL GENETICS

SECOND EDITION

A. W. F. EDWARDS

Reader in Biometry
University of Cambridge

CAMBRIDGE
UNIVERSITY PRESS

CAMBRIDGE UNIVERSITY PRESS
Cambridge, New York, Melbourne, Madrid, Cape Town,
Singapore, São Paulo, Delhi, Tokyo, Mexico City

Cambridge University Press
The Edinburgh Building, Cambridge CB2 8RU, UK

Published in the United States of America by Cambridge University Press, New York

www.cambridge.org
Information on this title: www.cambridge.org/9780521775441

First edition published 1977
Second edition published 2000

A catalogue record for this publication is available from the British Library

ISBN 978-0-521-77544-1 Paperback

This edition is dedicated to

C.A.B. Smith

Weldon Professor of Biometry
University College London, 1964–1982

CONTENTS

PREFACE TO THE
SECOND EDITION

This new edition of *Foundations of Mathematical Genetics* affords an opportunity to add a chapter on Fisher's Fundamental Theorem of Natural Selection (chapter 9). Advances in understanding in the past twenty-five years have finally clarified this topic, which was omitted because of its uncertain interpretation when the book was first published in 1977. In other respects the text is mostly unchanged, for it is still intended to meet its original objective of presenting the elementary mathematical theory of genetics 'in a canonical form as a piece of mathematics standing in its own right', and little else has been added to that theory over the intervening years. The new chapter 9 replaces an epilogue which contained some contemporary comments of passing interest. An alternative proof of Feller's Theorem (section 5.2) with an extended discussion may be found in Edwards (1977) but has not been incorporated.

In at least one respect the terminology itself has changed, however. *Genetic variance* is now universally called *genic variance*. In Fisher's original terminology the *genetic variance* was that part of the total variance due to the additive effects of the genes, *genetic* being the adjective from *gene*, whereas the total variance, due to the genotypes, became the *genotypic variance*, *genotypic* being the adjective from *genotype*. Such linguistic logic has proved unsustainable, and the reader must interpret *genetic variance* in this book as the modern *genic variance*.

Six special copies of the 1977 edition were printed (as an unsuccessful submission for the 1975–76 Adams Prize of the University of Cambridge) with a different preface which might be of interest to historians. It finally bore fruit in my Fisher Memorial Lecture (Edwards, 1995). Cambridge University Library and Gonville and Caius College Library both possess copies.

It is a pleasure to add to the names of those whose help I acknowledged in the original preface that of Professor Warren

Ewens, with whom I have enjoyed discussing Fisher's Fundamental Theorem for more than twenty years.

A.W.F.Edwards

Gonville and Caius College
Cambridge *May 1999*

PREFACE TO THE
FIRST EDITION

'I believe that no one who is familiar, either with mathematical advances in other fields, or with the range of special biological conditions to be considered, would ever conceive that everything could be summed up in a single mathematical formula, however complex.'

R. A. Fisher (*1932*)

For many years it has seemed to me that there has been a need for an exposition of the basic mathematics of population genetics. The recent appearance of many excellent texts in mathematical population genetics has increased, rather than reduced, this need, for they tend to stress the advanced theory at the expense of the elementary, or the biological applications at the expense of the mathematical details.

To suggest that a purely mathematical account of the foundations may be timely is not to deny the importance of either the advanced theory or the biological applications; it is merely to recognize that the elementary mathematical theory has reached a sufficient maturity to be presented in a canonical form as a piece of mathematics standing in its own right.

The subject matter divides conveniently into the deterministic, encompassing the theory of gene frequencies in populations assumed large enough for stochastic variation to be neglected, and into the stochastic itself, where such chance variation is considered. To some extent the stochastic theory is the child of the deterministic, which therefore merits earlier attention. The present volume treats only the deterministic theory; I leave the exposition of the stochastic theory for others, better qualified than I, to undertake.

I have remarked elsewhere (Edwards, 1967) that 'One of the inevitable corollaries of the fact that the structure of mathematical genetics was created by three men of exceptional calibre, R. A. Fisher, J. B. S. Haldane and S. Wright, has been that the ground floor of the edifice is scantily furnished.' They ascended rapidly to the upper storeys, not always by the same staircase. The present account seeks to furnish the ground floor in such a way that, though the reader may query the style, selection, or arrangement, of the pieces, he will, I trust, find no fault with the wood or the glue. The style is purely mathematical for the reason already given, and uses such concepts (matrices, for example) as seem appropriate for the optimum treatment of each problem. The topics are selected from the theory of random-mating diploid populations with constant viabilities because this theory is the most developed (on account of its relevance to man and other higher organisms), and the selection of treatments reflects my firmly held opinion that some of the treatments most often presented conceal more than they reveal.

The notation is explained as the argument proceeds, starting with the description of a simple 'mathematical' organism in chapter 1. The symbol ** is used to indicate the end of a proof, or the end of a theorem given without explicit proof.

Much of chapter 3 derives from some joint work with Dr Chris Cannings, now of the University of Sheffield; his Ph.D. dissertation (1968 a) is full of useful comments on many of the models we consider. The program for drawing figure 3.3 was written by Mrs Agnes Corfield, now of the University of Kent, and executed on the computer of the University of St Andrews by courtesy of Professor A. J. Cole. In drafting chapter 4 I was helped by correspondence with Professor John Kingman, University of Oxford, and Dr Eugene Seneta, Australian National University; the matrix form of Kingman's proof of theorem 4.3.2 is due to Dr Jim Roger, now of the University of Reading. The figures in chapter 5 showing mappings of the gene-frequency space were plotted in the Computer Laboratory of the University of Cambridge, using a program originally devised at my suggestion by Mrs Corfield and Dr Ann Eyland, now of Macquarie University, Sydney. I am indebted to Mr Rudolph Hanka, of this Department, for helping me with this and other computing matters.

Above all, however, I am indebted to three former students of mine, Dr Cannings, Dr Eyland, and Dr Elizabeth Thompson, Fellow of King's College, Cambridge, for commenting on chapter drafts and for sustaining my interest at a time when other academic preoccupations loomed large.

A. W. F. Edwards

University Department of Community Medicine
Fenner's, Cambridge *September 1976*

CHAPTER I

THE GENETIC MODEL

I.I MENDEL'S FIRST LAW

Readers who are familiar with the biological basis of heredity will not need an introduction to the technical language by which it is customarily described, whilst experience shows that others (such as mathematicians with no previous biological interest) find the definition of technical words in terms of cellular structure and function a rich source of confusion. For the latter group we recommend a quick excursion into the popular genetics literature – an encyclopaedia article will probably suffice – because our plan is to introduce only those words necessary for a description of the mathematical operation of the genetic model, and to attach to each only its operational meaning in terms of that model. Readers already familiar with the terms will then be able to relate the mathematical function to the underlying biological mechanism, whilst others (if the attempt is successful) will find a self-contained account of a purely mathematical model whose consequences are explored in the rest of the book.

We start with the concept of a *locus* which may initially be thought of as a box with a capacity of two *genes*. Each of the multitude of different kinds of genes has its own specific locus, and all the genes which relate to a specific locus are known as the *alleles* for that locus.† Thus a locus labelled A may contain any two of the genes $a_1, a_2, ..., a_k$, including two the same, and k may be (in principle) any positive integer, specific to each locus. If $k = 1$ there is no variation, and the case is of no interest; if $k = 2$ the locus is said to be *diallelic*,

---

† There is an increasing tendency to restrict the use of the word 'gene' so that two genes are only said to be the same if they are identical by descent, the one being an exact copy of the other, and to refer to two genes which, though not necessarily identical by descent, are identical in effect, as being the same 'allele'. We shall, however, continue the customary usage, which is common throughout the literature on which this book depends, and write of 'gene frequencies' and 'genotypes' rather than 'allele frequencies' and 'allelotypes'.

1

and if more, *multiallelic*. Each individual possesses one locus of each kind, $A, B, C, ...$, and each locus is associated with a particular *chromosome*. However, this need not yet concern us, because for the first five chapters we shall only consider a single locus (two are considered in chapter 8) which is not associated with the X-*chromosome* (a chromosome involved in the determination of sex, which we consider in chapter 6) but with one of the other chromosomes, called *autosomes*.

The *genotype* of an individual at a single locus is specified by the pair of genes he carries at that locus. If the two members are the same (such as $a_1 a_1$) he is a *homozygote*, and otherwise a *heterozygote*, for that locus. The totality of genotypes, over all loci, is the overall genotype of the individual. At a single locus there will be k possible homozygous genotypes and $\frac{1}{2}k(k-1)$ possible heterozygous genotypes (since order is irrelevant), making $\frac{1}{2}k(k+1)$ in all. But in practice it may not be possible to distinguish, at the level of observation, each of these; what is observed is called the *phenotype*, and each phenotype may correspond to one or more genotypes. The number of different ways in which genotypes can be classified into phenotypes is a complicated problem involving the theory of partitions, and need not concern us. The simplest way in which two genotypes give the same phenotype is when $a_i a_i$ is indistinguishable from $a_i a_j$, in which case gene a_i is said to be *dominant* and gene a_j *recessive*, each with respect to the other. If the phenotypic expression is metrical, then two alleles are said to exhibit 'no dominance in an additive sense' when the heterozygote's measure is the arithmetic mean of the corresponding homozygotes', and 'no dominance in a multiplicative sense' when the measure is the geometric mean. Just as with a single locus, the individual's overall phenotype depends on his overall genotype, and not all genotypes may be distinguishable.

Reproduction is a two-stage process. In the first stage, each parent produces a *gamete* (the sperm of the male, and the egg of the female). Each gamete carries a complete set of chromosomes, consisting of one copy of each autosome and a sex chromosome (whose mode of inheritance is treated in chapter 6); the precise constitution of this chromosomal set need not be considered if we are treating a single locus, because in this case the only material fact is covered by:

2

Mendel's First Law (Mendel, 1866)

Each of the two genes at a single locus has a probability of one half of being the single gene at that locus carried by a particular gamete (whether maternal or paternal). **

The second stage of the process is that of fertilization, at which the egg and sperm unite to form a new individual, thus restoring two genes to each locus. The new cell carries two complete sets of chromosomes, and from it, by cell-division, the individual will grow. Gametes, carrying one set, are called *haploid*, while individuals are called *diploid*. We see how our image of a locus being, in an individual, a box with space for two genes has as its physical basis a pair of chromosomes. Man has 22 pairs of autosomes and one pair of sex chromosomes.

Mendel's Second Law, of independent segregation for two loci, will be introduced in chapter 8. The next section summarizes the salient characteristics of the 'organism' whose population genetics we shall investigate.

I.2   A SIMPLE ORGANISM

We base our mathematical account on a bisexual diploid organism of sufficient complexity to allow for the development of the theory, but otherwise as simple as possible. We endow it with some pairs of autosomes and a pair of sex chromosomes. A and B are autosomal loci with alleles $a_1, a_2, ..., a_k$ and b_1, b_2 respectively; sometimes k will be 2, sometimes 3, and sometimes arbitrary. A and B are linked with recombination fraction r, though we will sometimes allow $r = \frac{1}{2}$ so as to include the unlinked case (when A and B might be on different chromosomes, or *asyntenic* (Renwick, 1969)).

The sex chromosomes will be denoted by X and Y, XX being female and XY male. X carries a single locus (which we shall call X when there is no risk of confusion) with alleles $x_1, x_2, ..., x_k$, where k will sometimes be 2, and sometimes arbitrary.

All genes will be assumed stable, and mutation will not be taken into account.

The life cycle of the organism is of extreme simplicity. All individuals are of the same age. Mating takes place when they reach a certain age, and the proportions of the various genotypes amongst

3

the offspring are determined by the products of the proportions of the various gametes in the male and female parts of the population. This rule, known as the 'random union of gametes', is, under certain conditions which will be mentioned below, a result of mating taking place 'at random', that is, without reference to genotype. The gametes are produced in accordance with Mendel's First Law, and hence the proportion of maternal gametes carrying a given gene is equal to the proportion of genes of that kind amongst the females, and similarly for the males. When two loci are involved, if they are asyntenic Mendel's Second Law applies and the proportions of the various gametes will be given by the products of the gene proportions for the individual loci, but if they are syntenic and linked the segregations are no longer independent and the proportions will depend upon the recombination fraction in a way to be described. After reproduction all the adult individuals die, and the population consists solely of the offspring. Such a life-cycle is said to involve 'discrete non-overlapping generations'.

The population is supposed indefinitely large, and we will only concern ourselves with *proportions*, whether of genes, gametes, genotypes, or phenotypes. There is a long-established practice of referring to such proportions as 'frequencies', which we will not attempt to reverse: thus we may refer to a 'gene frequency', meaning the proportion of genes of a particular kind.

A major interest will be the behaviour of gene frequencies under selection. The only aspect of selection we shall consider is where the proportion of each genotype that survives from birth to the age of mating differs from genotype to genotype, being given by a constant specific to each genotype. The survival proportion is a kind of 'selective coefficient', which we will refer to as a *viability*. Since only relative viabilities are relevant to the models we consider, we shall not adopt an upper limit of unity to the value of a viability: any non-negative number may be encountered, and the word 'relative' is to be understood. Thus throughout the book the models are invariant with respect to multiplication of the viabilities by a positive constant.

1.3 RANDOM MATING AND RANDOM UNION OF GAMETES

The assumption of 'random union of gametes' is of great convenience to the development of the theory, but we should be clear under what conditions it holds.

Definition. Random mating

If a population contains the genotypes $G_1, G_2, ..., G_l$ in proportions $m_1, m_2, ..., m_l$ amongst the males ($\Sigma m = 1$) and in proportions $f_1, f_2, ..., f_l$ amongst the females ($\Sigma f = 1$), and the proportion of matings of the type G_i (male) $\times G_j$ (female) is $m_i f_j$, the condition of *random mating* is said to be fulfilled. **

Definition. Random union of gametes

If a population produces the gametes $a_1, a_2, ..., a_k$ in proportions $s_1, s_2, ..., s_k$ amongst the males ($\Sigma s = 1$) and in proportions $t_1, t_2, ..., t_k$ amongst the females ($\Sigma t = 1$), each male individual being supposed to produce the same number of gametes, and similarly each female, and if the proportion of offspring of (ordered) genotype $a_x a_y$ is $s_x t_y$, the condition of *random union of gametes* is said to be fulfilled. **

Theorem 1.3.1. *Equivalence of random mating and random union of gametes*

Provided each mating produces the same number of offspring and the condition of random mating obtains, the proportions of genotypes amongst the offspring will be those given on the assumption of the random union of gametes.

Proof. Let genotype G_i produce gametes $a_1, a_2, ..., a_k$ in the proportions $p_{i1}, p_{i2}, ..., p_{ik}$. Since G_i will typically be $a_x a_y$, if $x \neq y$ then $p_{ix} = p_{iy} = \frac{1}{2}$ and all other ps are zero, whereas if $x = y$, $p_{ix} = 1$, by Mendel's First Law (section 1.1). Then in the mating G_i (male) $\times G_j$ (female) the probability of gametes a_x from the male and a_y from the female coming together is $p_{ix} p_{jy}$, and the probability of this mating is $m_i f_j$. The total genotypic array (or distribution of genotypes) amongst the offspring is therefore

$$\sum_{i,\,j} \{m_i f_j \sum_{x,\,y} p_{ix} p_{jy} a_x a_y\}, \tag{1.3.1}$$

in which the coefficient of $a_x a_y$ is the probability of the ordered genotype $a_x a_y$. (We adopt the convention that the gene derived from the father is written first.)

5

But (1.3.1) may be rewritten

$$\sum_{x,\,y} \{\sum_{i,\,j} m_i f_j p_{ix} p_{jy}\, a_x a_y\}$$

$$= \sum_{x,\,y} \{\sum_i m_i p_{ix}\, a_x\}\{\sum_j f_j p_{jy}\, a_y\}$$

$$= \{\sum_x \sum_i m_i p_{ix}\, a_x\}\{\sum_y \sum_j f_j p_{jy}\, a_y\}. \qquad (1.3.2)$$

Now $\sum_i m_i p_{ix}$ is the proportion of gametes a_x amongst the males, or s_x, and $\sum_j f_j p_{jy}$ the proportion of gametes a_y amongst the females, or t_y. Hence (1.3.2) may be written

$$\sum_x s_x a_x \cdot \sum_y t_y a_y, \qquad (1.3.3)$$

and the probability of the genotype $a_x a_y$ is thus $s_x t_y$, as is given directly by the assumption of the random union of gametes. **

We note that the theorem holds even if male and female genotypes produce gametes of the varying kinds in different proportions, because p_{jy} may be replaced by p'_{jy} throughout the proof, the prime referring to the probabilities of the maternal gametes. This would occur under gametic selection with viabilities different amongst the paternal and maternal gametes.

If mating is at random and the offspring of each mating equally numerous, the genotypic array of the offspring may thus be found by multiplying together the male and female gametic arrays of the parental generation, those arrays being found on the assumption that each individual contributes equally to the total gametic output of his sex. Theorem 1.3.1 has important consequences. It enables us to follow the course of gene-frequency change over the generations by considering gamete frequencies rather than genotype frequencies, and it leads to the *Hardy–Weinberg Theorem*.

1.4 THE HARDY–WEINBERG THEOREM

In this section we assume that the male and female arrays, both gametic and genotypic, are identical; henceforth genotypes will be unordered. Thus

$$s_x = t_x = p_x \text{ (say)}, \quad \text{for all } x.$$

Theorem 1.4.1. *The Hardy–Weinberg Theorem*

After one generation of random mating the genotypic array will be of the form

$$\{\textstyle\sum_{x} p_x a_x\}^2 \tag{1.4.1}$$

and will remain the same in subsequent generations.

Proof. The proof follows immediately by setting $s_x = t_x = p_x$ in (1.3.3) and then observing that a genotypic array $\{\sum_{x} p_x a_x\}^2$ implies, in turn, a gametic array $\sum_{x} p_x a_x$. **

The theorem derives its name from simultaneous and independent treatments of the diallelic case in 1908 by the Cambridge mathematician G. H. Hardy and the Stuttgart physician W. Weinberg.

A population whose genotype proportions conform to (1.4.1) is said to obey the *Hardy–Weinberg Law*, and to exhibit genotypes in *Hardy–Weinberg proportions*. It is also said to be in *Hardy–Weinberg equilibrium*, equilibrium here referring to the fact that there is no tendency for the variation caused by the co-existence of different genotypes to disappear. This ability to maintain genetic variation is one of the most important aspects of Mendelian genetics; it was lacking in earlier 'blending' theories of inheritance. Even after Mendel's laws were widely known, until the work of Hardy and Weinberg it was not generally understood that the frequency of a recessive gene would not decline in the absence of selection.

A necessary and sufficient condition for Hardy–Weinberg equilibrium is that the frequency of each heterozygote is given by twice the geometric mean of the frequencies of the two corresponding homozygotes, as is immediately evident from (1.4.1).

CHAPTER 2

TWO ALLELES AT A SINGLE LOCUS

2.1 NO SELECTION

We consider two alleles, a_1 and a_2, with initial frequencies p_m, q_m in the males, and p_f, q_f in the females. In view of theorem 1.3.1 this information is sufficient to enable us to write down the offspring genotype distribution immediately, whatever the parental genotype distributions (subject to the above gene frequencies). Henceforth, genotypes are unordered ($a_2 a_1 \equiv a_1 a_2$).

We find the offspring genotypic array to be:

$$\left.\begin{array}{ll} a_1 a_1 & p_m p_f \\ a_1 a_2 & p_m q_f + q_m p_f \\ a_2 a_2 & q_m q_f \end{array}\right\}. \tag{2.1.1}$$

The frequency of the gene a_1 amongst the offspring is

$$p_m p_f + \tfrac{1}{2}(p_m q_f + q_m p_f)$$
$$= \tfrac{1}{2}(p_m + p_f),$$

in accordance with the fact that the males and the females each contribute half of the genes of the offspring generation.

Writing
$$p = \tfrac{1}{2}(p_m + p_f), \quad q = \tfrac{1}{2}(q_m + q_f),$$

and assuming no association between the sex of an offspring and its genotype, the Hardy–Weinberg Theorem ensures that (as is entirely obvious with only two alleles) in the next generation, and thereafter; the genotype distribution is

$$\left.\begin{array}{ll} a_1 a_1 & p^2 \\ a_1 a_2 & 2pq \\ a_2 a_2 & q^2 \end{array}\right\}. \tag{2.1.2}$$

The original genotypic proportions (2.1.1) will only be Hardy–

8

Weinberg proportions if

$$(p_m q_f + q_m p_f)^2 = 4p_m p_f q_m q_f. \tag{2.1.3}$$

But the theorem on the inequality of arithmetic and geometric means tells us that

$$\left(\frac{p_m q_f + q_m p_f}{2}\right)^2 \geqslant p_m q_f q_m p_f, \tag{2.1.4}$$

with equality if and only if $p_m q_f = q_m p_f$, which is $p_m = p_f$. Thus the effect of any difference between the sexes in gene frequency is to increase the proportion of heterozygotes above the Hardy–Weinberg proportion for a population with gene frequency $p = \frac{1}{2}(p_m + p_f)$.

Henceforth we assume that the genotype frequencies are the same in both sexes, unless otherwise stated.

2.2 SELECTION WITH CONSTANT VIABILITIES

Let the genotypes $a_1 a_1$, $a_1 a_2$ and $a_2 a_2$ have viabilities w_{11}, w_{12} and w_{22} respectively, and let the gametes that form the first generation of offspring have frequencies $p a_1$ and $q a_2$. Then the offspring genotypic frequencies, before selection, will be $p^2 a_1 a_1$, $2pq a_1 a_2$ and $q^2 a_2 a_2$, and the *mean viability* of the offspring will be

$$w = p^2 w_{11} + 2pq w_{12} + q^2 w_{22},$$

which will also be the factor by which the population size is reduced through the action of selection if the viabilities are absolute and not relative. The adult genotypic frequencies will therefore be

$$\left. \begin{array}{ll} a_1 a_1 & p^2 w_{11}/w \\[4pt] a_1 a_2 & 2pq w_{12}/w \\[4pt] a_2 a_2 & q^2 w_{22}/w \end{array} \right\} \tag{2.2.1}$$

and the new gametic frequencies

$$\left. \begin{array}{l} p' = \dfrac{p^2 w_{11} + pq w_{12}}{w} \\[14pt] q' = \dfrac{pq w_{12} + q^2 w_{22}}{w} \end{array} \right\}. \tag{2.2.2}$$

It is more convenient to continue in terms of the *gene ratios* $u = p/q$, $u' = p'/q'$, for then (2.2.2) becomes

$$u' = u\,\frac{u w_{11} + w_{12}}{u w_{12} + w_{22}}. \tag{2.2.3}$$

We note that if u is positive, so is u', and hence so are all subsequent values. At gene-frequency equilibrium $u' = u = \hat{u}$ (say) and

$$\hat{u}w_{11} + w_{12} = \hat{u}w_{12} + w_{22}; \quad \text{or } \hat{u} = 0; \quad \text{or } \hat{u} = \infty.$$

Thus
$$\hat{u} = \frac{w_{12} - w_{22}}{w_{12} - w_{11}} \tag{2.2.4}$$

is an equilibrium. Since $0 \leqslant u \leqslant \infty$, there can only be equilibria in the permitted range $0 \leqslant p \leqslant 1$ if $(w_{12} - w_{22})$ and $(w_{12} - w_{11})$ are the same sign. The internal equilibrium value $\hat{p}$ is then

$$\frac{\hat{u}}{1 + \hat{u}} = \frac{w_{12} - w_{22}}{2w_{12} - w_{11} - w_{22}}. \tag{2.2.5}$$

This model was first given by R. A. Fisher in 1922, and he stated that 'if selection favours the heterozygote, there is a condition of stable equilibrium'.

Theorem 2.2.1. Fisher's Theorem

In the single-locus two-allele model the equilibrium gene ratios are

$$0, \infty, \text{ and } \hat{u} = \frac{w_{12} - w_{22}}{w_{12} - w_{11}}.$$

(Note that $\hat{u}$ is only a gene ratio if $(w_{12} - w_{22})$ and $(w_{12} - w_{11})$ are of the same sign; if either is zero $\hat{u}$ reduces to 0 or ∞; if both are zero every gene ratio is an equilibrium one.)

If $(w_{12} - w_{22})$ and $(w_{12} - w_{11})$ are both positive, $\hat{u}$ is a stable equilibrium, and the gene ratio will converge monotonically to it; if they are both negative $\hat{u}$ is an unstable equilibrium and the gene ratio will diverge monotonically from it. When $\hat{u}$ is negative, and hence not a gene ratio, there are two cases:

(1) $(w_{12} - w_{22})$ negative and $(w_{12} - w_{11})$ positive, when the gene ratio converges monotonically to $u = 0$;

(2) *vice versa*, when the gene ratio increases monotonically and indefinitely.

Proof. The equilibrium has been derived above, (2.2.4). Equation (2.2.3) may be written

$$u' = u \frac{(uw_{11} + w_{22}) + (w_{12} - w_{22})}{(uw_{11} + w_{22}) + u(w_{12} - w_{11})}$$

or
$$\frac{u'}{u} = \frac{(uw_{11} + w_{22}) + \hat{u}(w_{12} - w_{11})}{(uw_{11} + w_{22}) + u(w_{12} - w_{11})}; \tag{2.2.6}$$

after a little algebra this reduces to

$$u' - \hat{u} = (u - \hat{u})\left(\frac{w_{22} + uw_{11}}{w_{22} + uw_{12}}\right) \tag{2.2.7}$$

provided $w_{11} \neq w_{12}$ (if $w_{11} = w_{12}$, $\hat{u}$ is infinite). We assume u to be positive initially, and hence that it is always positive. Now if $w_{11} < w_{12}$ and $w_{22} < w_{12}$, $\hat{u}$ is positive and (2.2.7) ensures that u' lies between u and $\hat{u}$, indicating monotonic convergence to $\hat{u}$; if $w_{11} > w_{12}$ and $w_{22} > w_{12}$, $\hat{u}$ is still positive, but (2.2.7) then ensures that u lies between u' and $\hat{u}$, indicating monotonic divergence from $\hat{u}$; if $w_{11} < w_{12} < w_{22}$, $\hat{u}$ is negative and (2.2.7) ensures that u' is less than u, indicating monotonic decrease to $u = 0$ (since the decrease only disappears at $u = 0$); and if $w_{11} > w_{12} > w_{22}$, $\hat{u}$ is negative and (2.2.7) ensures that u' exceeds u, indicating monotonic increase in u, with ever-increasing steps. Finally, if $w_{11} = w_{12}$ we work with $q/p = v$ instead of $p/q = u$, and the argument proceeds as before. **

This proof is a slight extension of that given by Fisher in 1930.

We will not consider the various cases which occur when the viabilities take on special values, except to note that should $w_{12}^2 = w_{11}w_{22}$, so that the viabilities are in geometric progression and are said to exhibit no dominance in a multiplicative sense, then (2.2.3) reduces to

$$u' = u\sqrt{\frac{w_{11}}{w_{22}}}, \tag{2.2.8}$$

and there can be no internal equilibrium. Gene-frequency change is then the same as it would be if the gametes (and not the genotypes) differed in viability, in the ratio $\sqrt{w_{11}} : \sqrt{w_{22}}$ for $a_1 : a_2$. Selection acting in this way is referred to as *gametic selection*. Joint gametic and genotypic selection can always be represented as genotypic selection by an appropriate modification of the genotypic viabilities.

Much of the mathematical interest of diploid population genetics arises precisely because, unless there is no dominance in a multiplicative sense, the effects of selection cannot be represented by attaching viabilities to genes rather than genotypes. In the next section we introduce a formulation which makes use of the concepts of the *average excess* and *average effect* of a gene, which in turn allows us to partition the total variance in viability, exhibited by a population, into a component which may be thought of as due to the additive effects of the genes on genotypic viabilities (the *genetic variance*)†and a residual which may be thought of as being due to the effect of

† See the preface.

dominance (the *dominance variance*). The formulation offers some conceptual and notational advantages which will be particularly useful when dealing with three or more alleles.

Finally, we note that the general solution to the recurrence relation (2.2.3) is not known, though some special cases yield solutions. Their investigation is left as an exercise to the reader.

2.3 PROPERTIES OF THE MEAN VIABILITY

In order to form a notational link with the case of three or more alleles to be treated in chapter 3, we revert to the use of gene frequencies, writing $p = p_1$ and $q = p_2$. The mean viability is then

$$
\begin{aligned}
w &= p_1^2 w_{11} + 2p_1 p_2 w_{12} + p_2^2 w_{22} \\
&= \sum_{i,j} p_i p_j w_{ij}.
\end{aligned} \tag{2.3.1}
$$

Since genotypes are unordered, $w_{ji} = w_{ij}$. Equation (2.2.2) may be written

$$
p_i' = \sum_j \frac{p_j w_{ij}}{w} p_i, \quad i = 1, 2, \tag{2.3.2}
$$

and the mean viability in the next generation is

$$
w' = \sum_{i,j} p_i' p_j' w_{ij}. \tag{2.3.3}
$$

We now introduce the *average excess* of a gene.

Definition. Average excess

With respect to any character, the average excess of a gene is the difference between the mean for that sub-population consisting of all the homozygotes and half the heterozygotes which carry that gene, and the mean of the whole population. ******

This definition follows Fisher (1930), but we will not use his notation, which was a for average excess and α for average effect (to be introduced below). As we prove in theorem 2.3.1, $a = \alpha$ in our application, and we will therefore use α for average excess and then prove that it is also the average effect.

The average excess of a_1 with respect to viability is

$$
\begin{aligned}
\alpha_1 &= \frac{p_1^2 w_{11} + p_1 p_2 w_{12}}{p_1^2 + p_1 p_2} - w \\
&= p_1 w_{11} + p_2 w_{12} - w.
\end{aligned}
$$

This form shows that under Hardy–Weinberg equilibrium an equivalent definition of the average excess of a gene is the mean deviation from the population mean of individuals which received that gene from one parent, the gene received from the other parent having come at random from the population. Generally,

$$\alpha_i = \sum_j p_j w_{ij} - w. \qquad (2.3.4)$$

The concept is useful, for with it we may write (2.3.2) as

$$p_i' = \frac{\alpha_i + w}{w} p_i. \qquad (2.3.5)$$

Now the *genetic variance* (or *additive genetic variance*) is defined (Fisher, 1918) as the variance removed by the regression of the number of genes of each kind in a genotype on the value (in this case viability) of that genotype. Thus if we adopt the linear model

$$w_{ij} = w + \alpha_i^* + \alpha_j^* + \delta_{ij} \qquad (2.3.6)$$

and find the least-squares values of the α_i^*, we will be able to find the genetic variance. The least-squares values are known as the *average effects* of the genes (Fisher, 1930; figure 2.1), and to find them we shall prove another theorem of Fisher's which holds under random mating:

Theorem 2.3.1. *Average effect equals average excess*

Under random mating, the average effect of each gene is equal to its average excess.

Proof. From (2.3.6), all we have to show is that setting

$$\alpha_i^* = \alpha_i = \sum_j p_j w_{ij} - w \qquad (2.3.4\ bis)$$

minimizes $\qquad S = \sum_{i,j} p_i p_j (w - w_{ij} + \alpha_i^* + \alpha_j^*)^2. \qquad (2.3.7)$

Now $\qquad \dfrac{1}{2}\dfrac{\partial S}{\partial \alpha_i^*} = 2\sum_j p_i p_j (w - w_{ij} + \alpha_i^* + \alpha_j^*)$

$$= 2p_i(w - \sum_j p_j w_{ij} + \alpha_i^* + \sum_j p_j \alpha_j^*).$$

If we put $\alpha_i^* = \alpha_i$, then by (2.3.4) both $w - \sum_j p_j w_{ij} + \alpha_i^*$ and

13

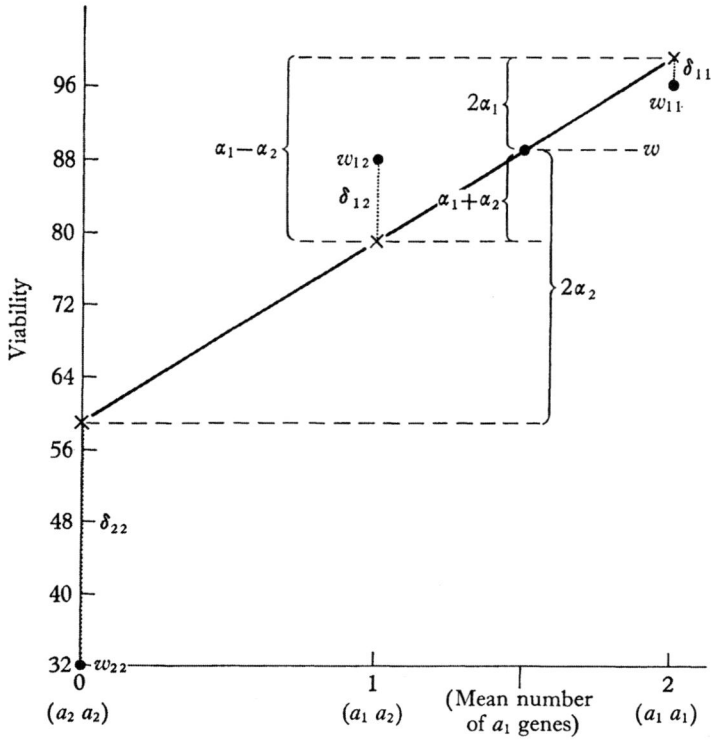

Figure 2.1. The linear model, (2.3.6). The line is the weighted least-squares regression of genotypic viability on the number of a_1 genes in the genotype, for viabilities $w_{11} = 96$, $w_{12} = 88$, $w_{22} = 32$ and gene frequency $p_1 = \frac{3}{4}$. The line passes through the overall mean ($w = 89$, mean number of a_1 genes per genotype $= 2p_1 = 1\frac{1}{2}$) and has gradient $\alpha_1 - \alpha_2 = 20$. Note the representation of the average effects ($\alpha_1 = 5$, $\alpha_2 = -15$) and the dominance deviations ($\delta_{11} = -3$, $\delta_{12} = 9$, $\delta_{22} = -27$). Redrawn from Falconer (1964), figure 7.2.

$\sum\limits_j p_j \alpha_j^*$ are zero. Thus $\alpha_i^* = \alpha_i$, all i, is the stationary value of S. That it is a minimum is obvious. **

The point of the theorem is that the genetic variance is defined in terms of the linear model, but the average excess is not. If, therefore, we wish to forge a link between the genetic variance and the change in mean viability (which is our intention), and if the former involves

14

the average effects and the latter the average excesses (as we shall see it does), we need to link the two concepts.

First, we derive the genetic variance. Thinking of the α_i as average effects, it must be

$$v = \sum_{i,j} p_i p_j (\alpha_i + \alpha_j)^2.$$

Now the weighted average of the average effects is zero (as is clear from (2.3.4)), or

$$\sum_i p_i \alpha_i = 0, \tag{2.3.8}$$

and its square must also be zero:

$$(\sum_i p_i \alpha_i)^2 = \sum_{i,j} p_i p_j \alpha_i \alpha_j = 0.$$

So the genetic variance

$$v = \sum_{i,j} p_i p_j (\alpha_i^2 + 2\alpha_i \alpha_j + \alpha_j^2)$$

may be written

$$v = 2\sum_{i,j} p_i p_j \alpha_i^2$$

$$= 2\sum_i p_i \alpha_i^2, \tag{2.3.9}$$

which is twice the variance of the average effects.

Now we derive an expression for the change in mean viability from one generation to the next, thinking of the α_i as average excesses. We write

$$\Delta p_i = p_i' - p_i,$$

whence from (2.3.5)

$$\Delta p_i = \frac{\alpha_i p_i}{w}, \tag{2.3.10}$$

and consider

$$\Delta w = w' - w$$

$$= \sum_{i,j} [(p_i + \Delta p_i)(p_j + \Delta p_j) - p_i p_j] w_{ij}$$

$$= 2\sum_{i,j} p_j \Delta p_i w_{ij} + \sum_{i,j} \Delta p_i \Delta p_j w_{ij}. \tag{2.3.11}$$

Now using (2.3.10) this becomes

$$\Delta w = 2\frac{\sum_{i,j} \alpha_i p_i p_j w_{ij}}{w} + \sum_{i,j} \Delta p_i \Delta p_j w_{ij}$$

15

and using (2.3.4) we have

$$\Delta w = 2 \frac{\sum\limits_{i} \alpha_i p_i (\alpha_i + w)}{w} + \sum\limits_{i,j} \Delta p_i \Delta p_j w_{ij}$$

$$= \frac{v}{w} + \sum\limits_{i,j} \Delta p_i \Delta p_j w_{ij}, \qquad (2.3.12)$$

by virtue of (2.3.8) and (2.3.9).

This form is due to Li (1969), after whom we name the theorem in section 4.3; for further comments see section 4.8 'Historical notes'.

We have used the summation notation even though treating the case of two alleles, since it conveys the structure of the argument more clearly. The interested reader may run through the development writing out the expressions fully for two alleles. Equation (2.3.12) may then be written as follows:

Theorem 2.3.2. Change in the mean viability

The change in the mean viability from one generation to the next at a diallelic locus is

$$\Delta w = \frac{v}{w} + (\Delta p_1)^2 (w_{11} - 2w_{12} + w_{22}) \qquad (2.3.13)$$

where v is the genetic variance in viability amongst the zygotes of the first generation, and $\Delta p_1 = -\Delta p_2$ is the change in gene frequency. **

We now reformulate (2.3.13) in order to demonstrate that Δw is non-negative. From (2.3.9) and (2.3.10) we find

$$(\Delta p_1)^2 = \frac{p_1 p_2 v}{2w^2}, \qquad (2.3.14)$$

whence (2.3.13) may be written

$$\Delta w = \frac{v}{2w^2} [2w + (w_{11} - 2w_{12} + w_{22}) p_1 p_2]$$

$$= \frac{v}{2w^2} (w + w_{11} p_1 + w_{22} p_2). \qquad (2.3.15)$$

16

Since this is manifestly non-negative we have the following:

Theorem 2.3.3. *Scheuer and Mandel's Theorem*

At a diallelic locus the mean viability never decreases from one generation to the next, and remains constant if and only if the gene frequencies are at their equilibrium values.

Proof. Equation (2.3.15) is non-negative; it is only zero when v is zero which, from (2.3.14), is only true at equilibrium. ∗∗

Scheuer and Mandel gave the above expression for Δw in 1959 (though they actually wrote out the genetic variance v in full in terms of the gene frequencies and viabilities; see also section 4.8 'Historical notes'). In point of fact their formulation implies a stronger inequality for Δw, because (2.3.15) may be written

$$\Delta w = \frac{v}{2w} + \frac{v}{2w^2}(w_{11}p_1 + w_{22}p_2), \qquad (2.3.16)$$

the second part of which is non-negative.

Theorem 2.3.4. $\Delta w \geqslant v/2w$

At a diallelic locus the increment in mean viability is greater than or equal to one half the genetic variance divided by the mean viability. ∗∗

As we shall see in chapter 4, theorem 2.3.3 generalizes to the multiallelic case, but not theorem 2.3.4.

Finally, we note that the last bracket of (2.3.13) may be written $(w_{11} - w_{12}) + (w_{22} - w_{12})$, so that from Fisher's Theorem (theorem 2.2.1) we see that if there is an internal stable equilibrium† this term is negative, but if there is an internal unstable equilibrium it is positive. Using (2.3.13) and theorem 2.3.4 we may therefore assert the following:

Theorem 2.3.5. *Inequalities for the change in mean viability*

When there is an internal stable equilibrium,

$$\frac{v}{w} \geqslant \Delta w \geqslant \frac{v}{2w},$$

† 'internal' excludes the boundary points $p = 0$ and $p = 1$.

but when there is an internal unstable equilibrium,

$$\Delta w \geqslant \frac{v}{w}. \quad **$$

Consider now the quadratic function w (2.3.1) in the gene-frequency space $p_1 + p_2 = 1$. The mean viability of a population not in equilibrium must necessarily increase in the next generation, by Scheuer and Mandel's Theorem, so that the points in the gene-frequency space representing successive generations must move 'up' the function w. Since a stable equilibrium may be approached monotonically from either side (theorem 2.2.1), it must correspond to a maximum of w, and since gene frequencies diverge monotonically from an unstable equilibrium, such an equilibrium must correspond to a minimum of w. Finally, if there is no equilibrium, w cannot possess a turning point in the permitted range. Moreover w, being a quadratic function, cannot have more than one turning point.

It thus follows that the behaviour of the function w may be used to determine the position and stability of the equilibrium.

Theorem 2.3.6. *The Equivalence Theorem*

The turning point of the function w is at the equilibrium gene frequency, and the equilibrium is stable if the turning point is a maximum, and unstable if it is a minimum. **

We would have like to have called this theorem after S. Wright, because it is essential to his formulation of the population genetics of a single locus (see Wright, 1969), but its explicit statement and proof are due to others. Of course, with only two alleles we do not need to use the increasing nature of w in the proof, because the theorem follows from the fact that the gene frequency changes monotonically and from a knowledge of the quadratic nature of w. But with more than two alleles one cannot appeal to monotonicity in the gene-frequency changes (Edwards, 1971).

In passing, we may note that the value of the mean viability amongst the offspring at equilibrium at a diallelic locus, found by inserting the equilibrium gene frequency (2.2.5) in the expression (2.3.1), is

$$\hat{w} = \frac{(w_{12}^2 - w_{11} w_{22})}{(2w_{12} - w_{11} - w_{22})}. \quad (2.3.17)$$

18

It should be noted that if $(w_{12} - w_{22})$ and $(w_{12} - w_{11})$ are of opposite sign, there is no equilibrium in the permitted range (Fisher's Theorem, theorem 2.2.1), and the above value refers to the unreal gene-frequency 'equilibrium', not to the value $p = 0$ or $p = 1$ to which the gene frequency will actually converge.

2.4   WRIGHT'S FORMULATION

The average excess (2.3.4) may be written

$$\alpha_i = \sum_j p_j w_{ij} - w = \frac{1}{2} \frac{\partial w}{\partial p_i} - w. \qquad (2.4.1)$$

We may thus write (2.3.5) as

$$p_i' = \frac{p_i}{2w} \cdot \frac{\partial w}{\partial p_i}. \qquad (2.4.2)$$

In this equation $\partial w / \partial p_i$ is the ordinary partial derivative of w with respect to p_i. It measures the gradient of the function w in the direction parallel to the ith axis. But of course gene frequencies are confined to the space $\Sigma p_i = 1$. Let us therefore consider (taking the case of two alleles) the gradient of w in the space $p_1 + p_2 = 1$, which may be taken as dw/dp_1 bearing in mind that $p_2 = 1 - p_1$.

$$dw = \frac{\partial w}{\partial p_1} dp_1 + \frac{\partial w}{\partial p_2} dp_2$$

and since $dp_2 = - dp_1$,

$$\frac{dw}{dp_1} = \frac{\partial w}{\partial p_1} - \frac{\partial w}{\partial p_2}. \qquad (2.4.3)$$

From (2.4.2)

$$\frac{\partial w}{\partial p_1} - \frac{\partial w}{\partial p_2} = 2w \left(\frac{p_1'}{p_1} - \frac{p_2'}{p_2} \right) = \frac{2w}{p_1 p_2} (p_1' p_2 - p_1 p_2') = \frac{2w \Delta p_1}{p_1 p_2},$$

whence from (2.4.3)

$$\Delta p_1 = \frac{p_1 p_2}{2w} \cdot \frac{dw}{dp_1}. \qquad (2.4.4)$$

This is Wright's equation (1937), and by the use of directional derivatives it may be extended to more than two alleles. It has the virtue of making clear the first part of the Equivalence Theorem (although in the next chapter we shall prove this for multiple alleles without using directional derivatives), but it has had the vice of

19

misleading investigators into thinking of w as a potential function†
in the gene-frequency space. We shall neither use it, nor pursue it
further, except to justify Fisher's statements (1941) that it amounts
to no more than saying that the average effect and the average excess
are the same under random mating, as recognized by Fisher in 1930
(p. 35). For the *average excess* led us to

$$\Delta p_i = \frac{\alpha_i \, p_i}{w} \qquad\qquad (2.3.10\,bis)$$

whilst the concept of the *average effect* implies

$$dw = 2\sum_i \alpha_i dp_i, \qquad\qquad (2.4.5)$$

as is clear from its original definition and least-squares derivation.
With two alleles we thus have

$$dw = 2(\alpha_1 - \alpha_2)\,dp_1$$

or
$$(\alpha_1 - \alpha_2) = \frac{1}{2}\frac{dw}{dp_1}. \qquad\qquad (2.4.6)$$

But interpreting the αs as average excesses, $(2.3.10\,bis)$ leads to

$$(\alpha_1 - \alpha_2) = w\left(\frac{\Delta p_1}{p_1} - \frac{\Delta p_2}{p_2}\right) = w\frac{\Delta p_1}{p_1 p_2} \qquad\qquad (2.4.7)$$

since
$$-\Delta p_2 = \Delta p_1.$$

If we now equate (2.4.6) and (2.4.7) we immediately recover (2.4.4).
The difference between the average effects, $(\alpha_1 - \alpha_2)$, is known as
the *average effect of the gene substitution* (figure 2.1).

Although we will not encounter (2.4.4) again, (2.4.2) occurs in
Baum and Eagon's Theorem (theorem 4.7.1).

---

† For further comments on Wright's formulation, see Edwards (1971,
2000).

CHAPTER 3

TWO ALLELES USING HOMOGENEOUS COORDINATES

3.1 GRAPHICAL REPRESENTATION

Since gene 'frequencies' and genotype 'frequencies' are in reality proportions, it is natural to enquire whether there is any advantage in using homogeneous coordinates to represent them. For example, the proportions of three genotypes in a population may be represented on a triangular diagram in which the coordinates of a point are proportional to the lengths of the perpendiculars to the three sides (de Finetti, 1926). Cannings and Edwards (1968) have pursued de Finetti's representation furthest, and this chapter is based on their account (see also Yaglom, 1967, and Ineichen and Batschelet, 1975).

It is convenient to use an equilateral triangle of unit height as the reference frame, for then the three perpendiculars l, m, n (figure 3.1) representing proportions sum to unity. Such a coordinate system is properly called *trilinear*; a related system is that of *areal* coordinates, in which the proportions are represented by the areas of the three triangles formed by the lines joining the given point to the vertices. When the reference triangle is equilateral of unit height the two systems are identical except for the constant factor $1/\sqrt{3}$, the area of the triangle. Both are examples of *homogeneous* coordinates, since any equation can be rendered homogeneous by using the fact that the sum of the coordinates is a constant. It is this homogeneity which is attractive, and we will usually adopt it. The principal use of homogeneous coordinates is in the algebraic representation of the results of projective geometry, where one is indifferent as to whether trilinear or areal coordinates are intended, and where the reference triangle is arbitrary. The present usage, however, is metrical, whence our adoption of a particular reference triangle. The reader is at liberty to think in terms of either trilinear or areal coordinates as he pleases, since with an equilateral reference triangle they differ only by a constant, and results will generally be quoted in homo-

21

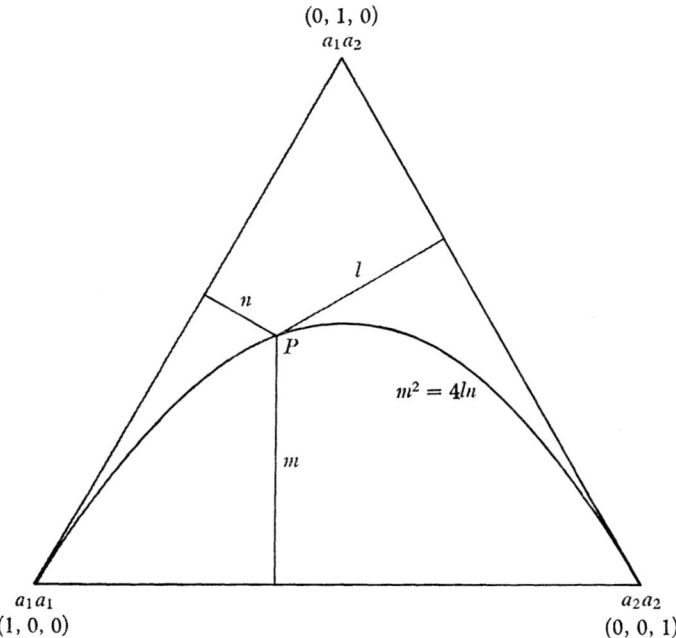

Figure 3.1. The loci of populations with genotypic proportions
in Hardy–Weinberg equilibrium.

geneous form in which the sum of the coordinates is irrelevant. For
this reason the coordinates of a point will not usually be normalized,
allowing the economy of writing, for example, (a, b, c) instead of
$(a/(a+b+c), b/(a+b+c), c/(a+b+c))$. Occasionally it will be necessary to introduce an inhomogeneous equation, such as that for the
mean viability

$$w = w_{11}\,l + w_{12}\,m + w_{22}\,n,$$

and the side equation $l+m+n = 1$ should then be borne in mind.
In chapter 5 it will be necessary, at a certain stage, to use a non-
equilateral reference triangle, and reasons will there be given for
adopting areal rather than trilinear coordinates.†

† For a modern account of the use of homogeneous coordinates, see *Algebraic
Projective Geometry* by J. G. Semple and G. T. Kneebone (Clarendon Press,
Oxford, 1952 and later); this has a lucid part 1 'The origins and development
of geometrical knowledge' in which the role of homogeneous coordinates is
discussed. But I have to confess that my own source-book for metrical results
using trilinear coordinates is a little older: *An Elementary Treatise on Trilinear
Coordinates* by The Rev. N. M. Ferrers, Fellow and Tutor of Gonville and
Caius College (Macmillan, London and Cambridge, second edition 1866).

Let the genotypic proportions be $l\,a_1a_1$, $m\,a_1a_2$ and $n\,a_2a_2$. If these are in Hardy–Weinberg equilibrium, $m^2 = 4ln$ (section 1.4), which is the equation of a parabola passing through the base vertices of the triangle and the point $(\frac{1}{4}, \frac{1}{2}, \frac{1}{4})$, being tangential to the sides at the vertices (figure 3.1). The Hardy–Weinberg parabola may be regarded as the 'pre-selection' curve since it reflects the genotypic proportions amongst the zygotes before these proportions have been modified by differential viability. The effect of selection is to transform the pre-selection point (l, m, n) into the 'post-selection' point $(w_{11}\,l, w_{12}\,m, w_{22}\,n)$, and since the former satisfies $m^2 = 4ln$ the latter must satisfy $w_{11}w_{22}m^2 = 4w_{12}^2\,ln$. This equation is evidently that of a conic symmetrical about the vertical axis of the diagram, $l = n$, and dependent on only one constant, $k = w_{12}^2/w_{11}w_{22}$. For $k = 0$ it degenerates into $m = 0$, the base of the triangle, and for $k = \infty$ into $ln = 0$, the other two sides of the triangle. By considering the intersection of the conic $m^2 = 4kln$ and the line at infinity $l + m + n = 0$ it is apparent that for other values of k the conic takes the following forms:

$$\left.\begin{array}{lll} k < 1, & w_{12}^2 < w_{11}w_{22}: & \text{ellipse} \\[2mm] k = 1, & w_{12}^2 = w_{11}w_{22}: & \text{parabola} \\[2mm] k > 1, & w_{12}^2 > w_{11}w_{22}: & \text{hyperbola} \end{array}\right\}. \qquad (3.1.1)$$

The ellipses lie below the parabola; the hyperbolae above.

In the special case in which there is no dominance in a multiplicative sense so that $w_{12}^2 = w_{11}w_{22}$ (section 2.2), the conic is simply the Hardy–Weinberg parabola. In the general case $k \neq 1$ the conic still touches the sides of the triangle at the base vertices, since putting either $l = 0$ or $n = 0$ leads to the double root $m^2 = 0$. It follows that all the conics of the family have double contact with each other, and that they cannot cross each other elsewhere.

If every point on the pre-selection curve is joined by a straight line to its corresponding point on the post-selection curve, then selection may be viewed graphically (figure 3.2). Starting at the point representing zygotic proportions before selection (P), the population will move to the corresponding post-selection point (Q). Mating will then reassort the genes so that Hardy–Weinberg equilibrium is restored without changing the gene frequency, the population thus moving vertically back to the Hardy–Weinberg parabola (P'). The cycle is then repeated, and the population follows a sawtoothed path to its equilibrium position.

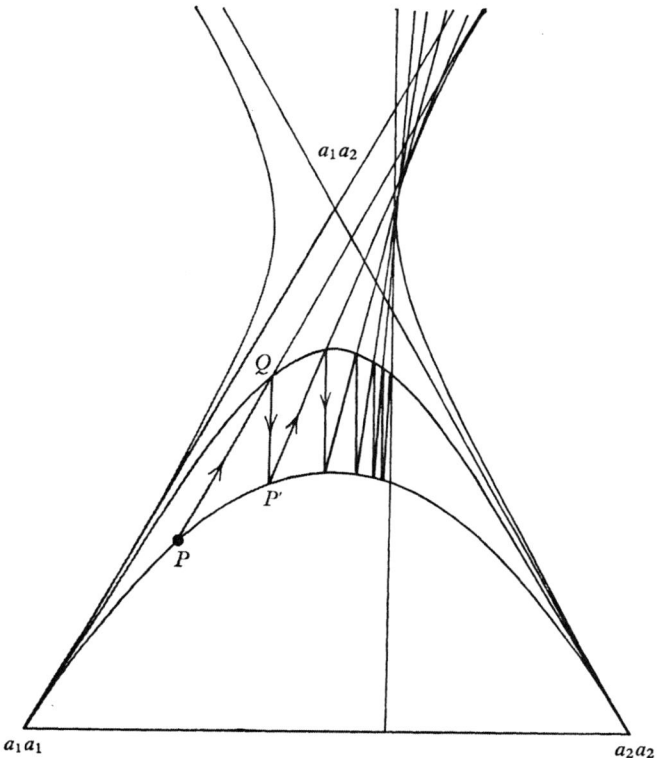

$a_1 a_2$

Q

P'

P

$a_1 a_1$ $a_2 a_2$

Figure 3.2. Selection viewed graphically. The curve through P and P' is the pre-selection curve, and that through Q the post-selection curve. The viabilities used in this example are ($\frac{1}{4}$, 1, $\frac{1}{2}$), the equilibrium gene frequency thus being 0.6 for the a_1 gene. See text for further explanation.

It is easy to see that vertical lines are lines of constant gene frequency, for the vertical displacement of the point (l, m, n) by an amount δ upwards leads to $(l-\frac{1}{2}\delta, m+\delta, n-\frac{1}{2}\delta)$ and the frequency of gene a_1 remains $p = l+\frac{1}{2}m$. Choosing $\delta = -m$ gives the intercept of the line with the base of the triangle as $(l+\frac{1}{2}m, 0, n+\frac{1}{2}m)$, the point which divides the base in the ratio $p:q$.

3.2 THE GENERAL DIAGRAM

In order to construct the lines joining corresponding points on the pre- and post-selection curves it is convenient to establish the equation for their envelope in the general case. The line joining (x, y, z)

to $(w_{11}x, w_{12}y, w_{22}z)$ has the equation

$$G = \begin{vmatrix} l & m & n \\ x & y & z \\ w_{11}x & w_{12}y & w_{22}z \end{vmatrix} = 0, \qquad (3.2.1)$$

and we know that $y^2 = 4xz$. Eliminating y and deferring, for the moment, the cases $x = 0$ and $z = 0$, we find

$$\begin{aligned} G' = G/\sqrt{(xz)} = &-2(w_{12}-w_{22})\,zl \\ &+ (w_{11}-w_{22})\sqrt{(xz)}\,m + 2(w_{12}-w_{11})\,xn = 0, \end{aligned}$$

whence the equation of the envelope is found by eliminating x and z between the equations

$$\partial G'/\partial x = \tfrac{1}{2}(w_{11}-w_{22})\sqrt{(z/x)}\,m + 2(w_{12}-w_{11})\,n = 0$$

and $\quad \partial G'/\partial z = -2(w_{12}-w_{22})\,l + \tfrac{1}{2}(w_{11}-w_{22})\sqrt{(x/z)}\,m = 0$

leading to $\qquad m^2 = \dfrac{-16(w_{12}-w_{11})(w_{12}-w_{22})\,ln}{(w_{11}-w_{22})^2}. \qquad (3.2.2)$

This is another member of the family of conics $m^2 = 4kln$. If $x = 0$ or $z = 0$ then $y = 0$, and these cases refer to the base vertices of the triangles which transform into themselves since the pre- and post-selection curves coincide at these points.

Thus the pre- and post-selection curves and the envelope of the joining lines are all members of the family

$$m^2 = 4kln.$$

$k = 1 \quad$ gives the pre-selection curve,

$k = w_{12}^2/w_{11}w_{22} \quad$ the post-selection curve,

and $\quad k = \dfrac{-4(w_{12}-w_{11})(w_{12}-w_{22})}{(w_{11}-w_{22})^2} \quad$ the envelope.

A diagram including curves for the complete range of values of k, at suitable intervals, may thus be used for any particular case, the appropriate post-selection curve and envelope being picked out and the course of selection charted from any starting point. Such a diagram is given in figure 3.3, and an example in figure 3.2.

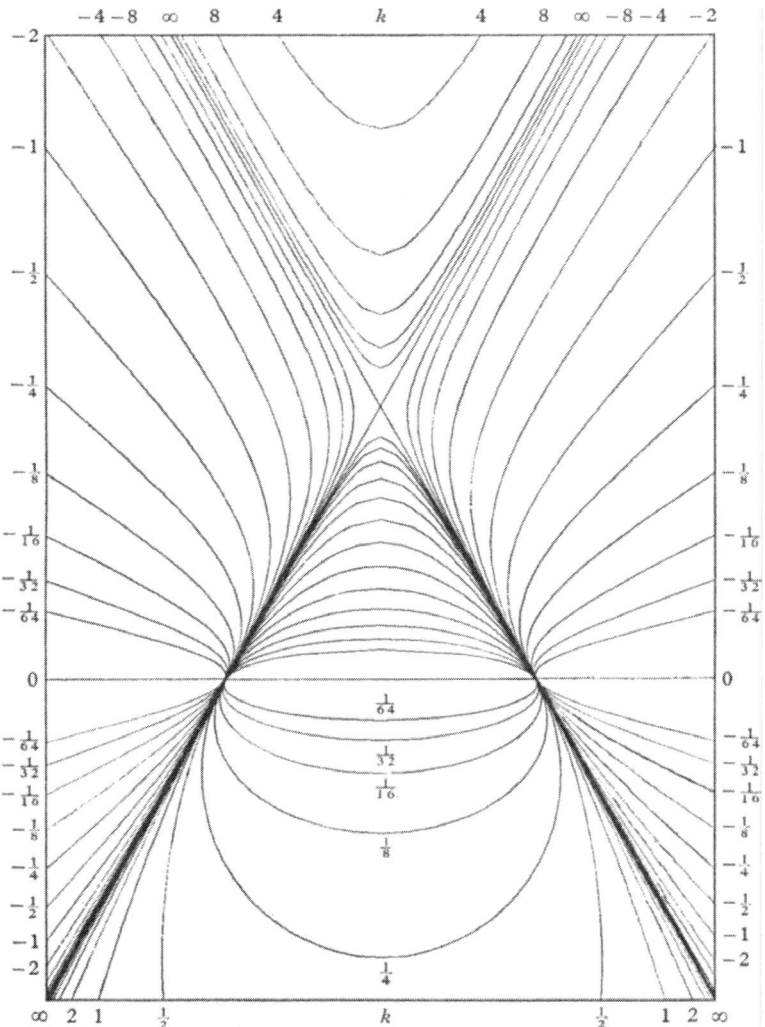

Figure 3.3. The general diagram of post-selection curves and envelopes, being the family of conics $m^2 = 4kln$.

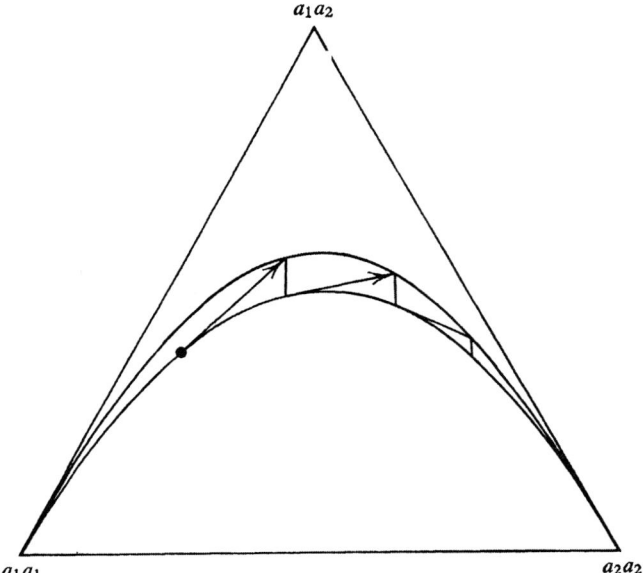

Figure 3.4. Selection with no dominance in viability in an additive sense, when the selection lines are tangential to the Hardy–Weinberg parabola: the case ($\frac{1}{5}$, $\frac{3}{5}$, 1).

3.3 SPECIAL CASES

The envelope (3.2.2) coincides with the Hardy–Weinberg parabola when

$$-4(w_{12}-w_{11})(w_{12}-w_{22}) = (w_{11}-w_{22})^2,$$

which is when

$$w_{11}+w_{22} = 2w_{12},$$

the case of no dominance in viability in an additive sense. The lines are then tangential to the Hardy–Weinberg parabola (figure 3.4).

When $w_{11} = w_{12}$ but $w_{22} \neq w_{12}$, as when a_1 is fully dominant to a_2, the equation for the line, (3.2.1), reveals that it always passes through the base vertex $a_2 a_2$ (0, 0, 1). The envelope (3.2.2) appears to be $m = 0$, but inspection of the preceding equations shows that in addition $l = 0$, and the envelope is thus indeed the single point (0, 0, 1), the family of lines being a pencil. Conversely, when $w_{22} = w_{12}$ but $w_{11} \neq w_{12}$, the lines form a pencil through (1, 0, 0) (figure 3.5).

When $w_{11} = w_{22} \neq w_{12}$, the symmetric case, the lines are found to form a pencil through the vertex $a_1 a_2$ (0, 1, 0) (figure 3.6), and

27

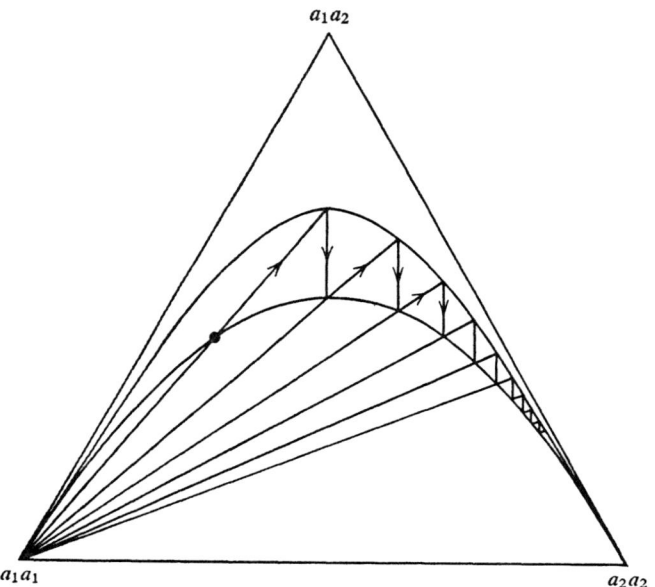

Figure 3.5. Selection with a_2 dominant to a_1, when the selection lines form a pencil through the a_1a_1 vertex: the case $(\frac{1}{4}, 1, 1)$.

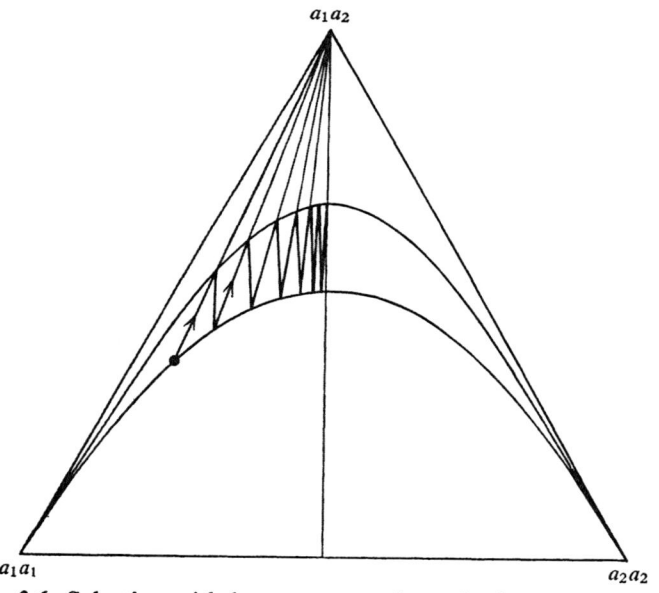

Figure 3.6. Selection with homozygotes of equal selective value, when the selection lines for a pencil through the a_1a_2 vertex: the case $(\frac{1}{2}, 1, \frac{1}{2})$. The equilibrium gene frequency is one half.

28

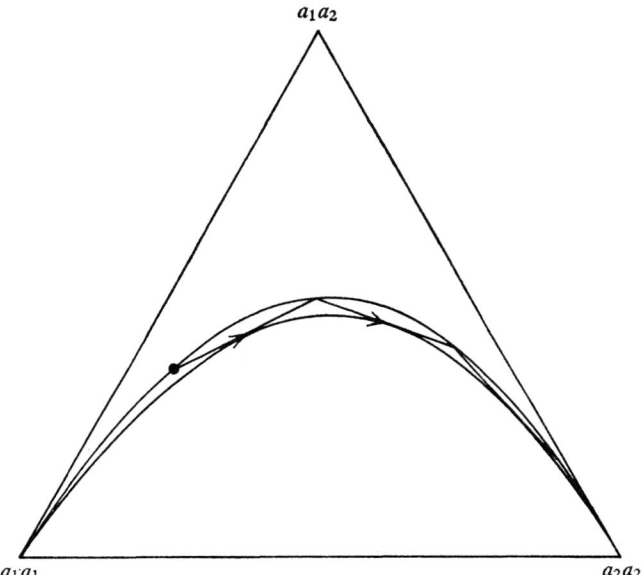

Figure 3.7. Selection with no dominance in viability in a multiplicative sense. The pre- and post-selection curves coincide, the envelope lying just below: the case $(\frac{1}{9}, \frac{1}{3}, 1)$.

when there is no dominance in viability in a multiplicative sense the envelope lies just below the Hardy–Weinberg parabola which, as we have seen (section 3.1), is also the post-selection curve (figure 3.7).

3.4   EQUILIBRIA, STABILITY AND CONVERGENCE

The equilibria of the system are of two kinds: those where the pre- and post-selection points coincide, in which case neither selection nor mating changes the genotype frequencies; and those where the selection line is vertical, in which case mating precisely restores the genotype frequencies to their pre-selection values, the gene frequency remaining unchanged.

Since the pre-selection point is $(p^2, 2pq, q^2)$ and the post-selection point is $(w_{11}p^2, 2w_{12}pq, w_{22}q^2)$, the former condition holds only when $w_{11} = w_{12} = w_{22}$, and there is no selection, or when $p = 0$ or when $q = 0$, the two boundary equilibria. The latter condition holds when the selection line intercepts the base in $(p, 0, q)$ (figure 3.8).

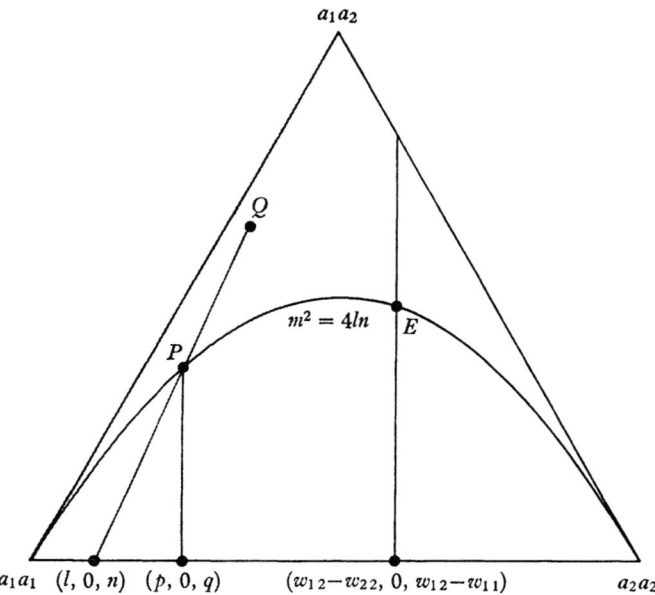

$a_1 a_2$

$m^2 = 4ln$ E

P

Q

$a_1 a_1$ $(l, 0, n)$ $(p, 0, q)$ $(w_{12}-w_{22}, 0, w_{12}-w_{11})$ $a_2 a_2$

Figure 3.8. The selection line PQ always leans towards the vertical through the equilibrium point E,

$$((w_{12}-w_{22})^2, 2(w_{12}-w_{22})(w_{12}-w_{11}), (w_{12}-w_{11})^2).$$

In general the intercept is $(l, 0, n)$ where

$$\begin{vmatrix} l & 0 & n \\ p^2 & 2pq & q^2 \\ w_{11}p^2 & 2w_{12}pq & w_{22}q^2 \end{vmatrix} = 0 \quad \text{and} \quad l+n = 1.$$

This reduces to

$$\frac{l}{n} = \frac{w_{12}-w_{11}}{w_{12}-w_{22}}\frac{p^2}{q^2} \tag{3.4.1}$$

and there will be equilibrium when $l/n = p/q$, in which case

$$\frac{p}{q} = \frac{w_{12}-w_{22}}{w_{12}-w_{11}},$$

in accordance with Fisher's Theorem (theorem 2.2.1). The equilibrium point on the Hardy–Weinberg parabola is thus

$$((w_{12}-w_{22})^2, 2(w_{12}-w_{22})(w_{12}-w_{11}), (w_{12}-w_{11})^2). \tag{3.4.2}$$

That the equilibrium will lie in the range $0 < p < 1$ only if $(w_{12}-w_{22})$ and $(w_{12}-w_{11})$ are of the same sign is obvious, since otherwise the equilibrium point (3.4.2) will lie outside the triangle of reference. We shall refer to an equilibrium in the range $0 < p < 1$ as *internal* and to one in the range $p < 0$ or $p > 1$ as *external*. Since the gene frequency cannot approach an external equilibrium point, but, when one exists, converges on $p = 0$ or $p = 1$, we call these points *boundary equilibria*. The coefficient in the equation for the envelope (3.2.2) must be negative for an internal equilibrium, showing that the envelope is a hyperbola, as in figure 3.2. It follows that the selection lines on the left of the equilibrium point have positive slope (in the normal Cartesian sense), and those on the right negative slope, and that the gene frequency always changes monotonically, since no two selection lines can intersect within the triangle.

Thus an internal equilibrium will be stable if the post-selection conic lies above the Hardy–Weinberg parabola, and unstable if it lies below. But these events have already (section 3.1) been seen to correspond to $w_{12}^2 > w_{11}w_{22}$ (hyperbola) and $w_{12}^2 < w_{11}w_{22}$ (ellipse) respectively, thus confirming the rules for determining the stability of an equilibrium established by Fisher's Theorem.

When there is no internal equilibrium the monotonicity of gene-frequency change is not so immediately obvious from the diagram, but since it has already been established, it will henceforth be assumed.

3.5   THE MEAN VIABILITY

The mean viability of a population with genotype frequencies l, m, n is

$$w = w_{11}l + w_{12}m + w_{22}n,$$

so that the locus of populations with mean viability w is

$$(w_{11} - w)l + (w_{12} - w)m + (w_{22} - w)n = 0, \qquad (3.5.1)$$

a straight line. Since moving from any point (l, m, n) to a new point $(l + \lambda, m + \mu, n + \nu)$, where $\lambda + \mu + \nu = 0$, changes w by an amount $(w_{11}\lambda + w_{12}\mu + w_{22}\nu)$ independently of (l, m, n), it is clear that for varying w (3.5.1) is a family of parallel lines which, when traversed in any specified direction, exhibit a constant rate of change of w. In other words, if w is represented by the dimension perpendicular to the plane of the reference triangle, the w-surface is itself a plane (inclined in general).

The mean viability amongst the offspring at equilibrium is

$$\hat{w} = \frac{w_{12}^2 - w_{11} w_{22}}{2w_{12} - w_{11} - w_{22}}, \qquad (2.3.17 \, bis)$$

and the line of constant viability through the equilibrium point therefore

$$w_{11} l + w_{12} m + w_{22} n = \frac{w_{12}^2 - w_{11} w_{22}}{2w_{12} - w_{11} - w_{22}}$$

which is

$$(w_{12} - w_{11})^2 l - (w_{12} - w_{11})(w_{12} - w_{22}) m + (w_{12} - w_{22})^2 n = 0. \qquad (3.5.2)$$

Solving this with $m^2 = 4ln$ gives the perfect square

$$[(w_{12} - w_{11})^2 l - (w_{12} - w_{22})^2 n]^2 = 0,$$

which indicates that (3.5.2) is tangent to the Hardy–Weinberg parabola at the internal or external equilibrium point (figure 3.9). Thus this point is a stationary value of w, the mean viability, in accordance with Scheuer and Mandel's Theorem (theorem 2.3.3).

This establishes the direction in which the lines of constant w lie, but not the sense in which the w-plane is inclined. For the latter we note that at the apex of the triangle, $(0, 1, 0)$, $w = w_{12}$ and at the midpoint of the base, $(\frac{1}{2}, 0, \frac{1}{2})$, $w = \frac{1}{2}(w_{11} + w_{22})$, from (3.5.1), whence the w-plane tilts up in a direction 'outwards' from the Hardy–Weinberg parabola at the equilibrium point if $2w_{12} > w_{11} + w_{22}$, and inwards if $2w_{12} < w_{11} + w_{22}$. Thus w decreases along the parabola in directions away from the equilibrium if $2w_{12} > w_{11} + w_{22}$ and in directions towards the equilibrium if $2w_{12} < w_{11} + w_{22}$. In the case of equality (no dominance in viability in an additive sense) the lines of constant w are parallel to the axis of the Hardy–Weinberg parabola, the inclination of the w-surface being down to the right if $w_{11} > w_{22}$, and otherwise to the left. In this case the mean viability of the offspring is equal to the mean viability of the adults of the previous generation: all populations with the same gene frequency have the same mean viability. This is not true in the other case in which the mean viability of the offspring equals the mean viability of the adults, namely when there is no dominance in a multiplicative sense. The equality then arises because random mating does not change the genotype frequencies at all (figure 3.7). It is interesting that the condition for the mean viability not to change from the population

32

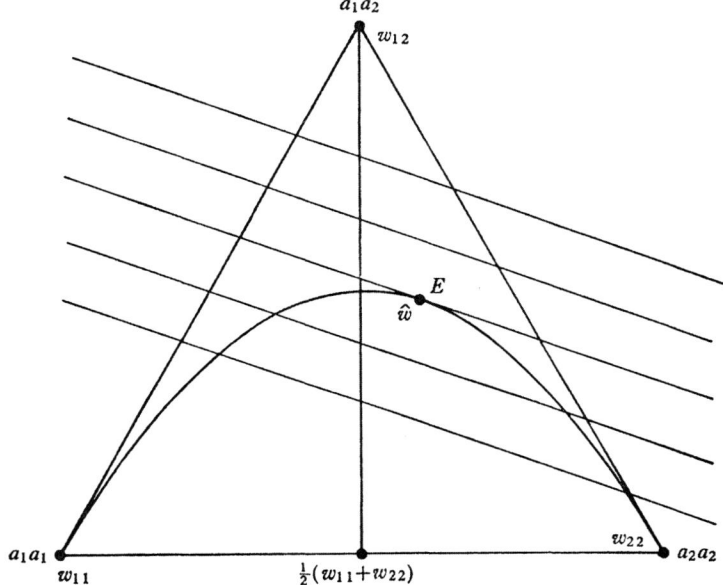

Figure 3.9. Lines of constant w, and values at specific points, for the case of an internal equilibrium point E.

of adults to the population of their offspring involves both additive and multiplicative kinds of lack of dominance.

It remains to demonstrate Scheuer and Mandel's Theorem (theorem 2.3.3), that the mean viability w increases from generation to generation except at equilibrium. We have seen that the lines of constant w lie in a direction tangential to the Hardy–Weinberg parabola at the equilibrium point, whether internal or external, and that w increases 'outwards' from the equilibrium if and only if $2w_{12} > w_{11} + w_{22}$. It is thus clear that, in view of the monotonic change in gene frequency, w always changes monotonically, and the only issue is whether in every case it increases:

1. w_{11}, $w_{22} < w_{12}$, whence $2w_{12} > w_{11} + w_{22}$. There is an internal stable equilibrium. Movement along the Hardy–Weinberg parabola is towards the equilibrium, in which direction w increases.

2. w_{11}, $w_{22} > w_{12}$, whence $2w_{12} < w_{11} + w_{22}$. There is an internal unstable equilibrium. Movement is away from the equilibrium, in which direction w increases.

33

3. $w_{11} < w_{12} < w_{22}$.

(i) $2w_{12} > w_{11} + w_{22}$, whence $(w_{12} - w_{11})^2 > (w_{12} - w_{22})^2$ and the equilibrium, which is external, must be to the left of $p = 1$ (see (3.4.2)). Thus w increases towards the right (p decreasing), which is the direction of gene-frequency change.

(ii) $2w_{12} < w_{11} + w_{22}$. By a similar argument the equilibrium is now external and to the right of $p = 0$, but now w increases towards it, to the right, which is still the direction of gene-frequency change.

4. $w_{11} > w_{12} > w_{22}$.

(i) $2w_{12} > w_{11} + w_{22}$, but now this implies $(w_{12} - w_{11})^2 < (w_{12} - w_{22})^2$ and the equilibrium is external and to the right of $p = 0$. Thus w increases to the left, which is the direction of gene-frequency change.

(ii) $2w_{12} < w_{11} + w_{22}$. A similar argument applies.

The special cases, when two viabilities are equal, present no problems, and the demonstration that the mean viability increases from generation to generation except at equilibrium is therefore complete.

CHAPTER 4

MANY ALLELES AT A SINGLE LOCUS

We consider k alleles, $a_1, a_2, ..., a_k$ with initial frequencies p_{m1}, p_{m2}, $...,p_{mk}$ in the males, and $p_{f1}, p_{f2}, ...,p_{fk}$ in the females. It is convenient to write these frequencies in the form of diagonal matrices:

$$P_m = \begin{pmatrix} p_{m1} & 0 & \cdots & 0 \\ 0 & p_{m2} & \cdots & 0 \\ \vdots & \vdots & & \vdots \\ 0 & 0 & \cdots & p_{mk} \end{pmatrix},$$

and similarly for P_f. Then in view of theorem 1.3.1 we may write down the offspring genotype distribution immediately as $P_m U P_f$ (where U is the $k \times k$ matrix with every element unity), the (i, j)th element being the frequency of the ordered genotype $a_i a_j$.

The frequency of the gene a_i amongst the offspring is thus

$$\tfrac{1}{2}\left(\sum_j p_{mi}p_{fj} + \sum_j p_{mj}p_{fi}\right) = \tfrac{1}{2}(p_{mi}+p_{fi}),$$

in accordance with the fact that the males and the females each contribute half of the genes of the offspring generation. In respect of any pair of alleles a similar result to that given in section 2.1 concerning an excess of heterozygotes holds.

Henceforth we assume that the genotype frequencies are the same in both sexes.

4.2 SELECTION WITH CONSTANT VIABILITIES

Let P be the $k \times k$ diagonal matrix of the frequencies of the genes $a_1, a_2, ..., a_k$, and W the symmetric matrix of genotypic viabilities in which w_{ij} $(= w_{ji})$ gives the viability of the genotype $a_i a_j$ $(\equiv a_j a_i)$. We assume that W is not singular, a restriction necessary in order to keep the treatment at the same level as the rest of the book. (We may note that for two alleles this restriction would eliminate from con-

35

sideration the case $w_{12}^2 = w_{11} w_{22}$, where there is no dominance in a multiplicative sense, and no internal equilibrium; see section 4.5.)

The array of genotypic frequencies before selection is

$$PUP,$$

in accordance with the Hardy–Weinberg Theorem (theorem 1.4.1), and after selection it will be simply proportional to

$$PWP. \qquad (4.2.1)$$

Theorem 4.2.1. *Mean viability*

The mean viability is $\quad w = 1^{\mathrm{T}}PWP1, \qquad (4.2.2)$

being the sum of the elements of (4.2.1). **

It follows that the normalized array of genotypic frequencies is

$$\frac{PWP}{w}, \qquad (4.2.3)$$

in an obvious notation. Following selection, the frequency of the ith gene is the sum of the elements of the ith row of (4.2.3), which we may write as:

Theorem 4.2.2. *Recurrence relation for the gene frequencies*

The gene frequencies in the next generation are given by

$$P'1 = \frac{PWP1}{w}. \qquad ** \qquad (4.2.4)$$

As with the case of two alleles (chapter 2) it is now convenient to introduce the *average effects* and the *genetic variance*. The former are defined by the weighted least-squares fit of the linear model, leading to

$$w_{ij} = w + \alpha_i + \alpha_j + \delta_{ij}, \qquad (4.2.5)$$

and:

Theorem 4.2.3. *The average effects*

If $\boldsymbol{\alpha}$ is the column vector of the average effects $\{\alpha_i\}$,

$$\boldsymbol{\alpha} = WP1 - w1. \qquad ** \qquad (4.2.6)$$

36

The proof parallels that of theorem 2.3.1, and is omitted. The genetic variance is defined to be the variance removed by the regression, and, as with two alleles, is twice the variance of the average effects:

$$v = 2\alpha^T P\alpha. \tag{4.2.7}$$

Theorem 4.2.4. *The change in gene frequency*

The change in gene frequency, represented by the diagonal matrix (ΔP), is given in terms of the average effects by

$$(\Delta P)1 = \frac{P\alpha}{w}. \tag{4.2.8}$$

Proof. Premultiplying (4.2.6) by P,

$$P\alpha = PWP1 - wP1,$$

whence, from (4.2.4), $wP'1 - wP1 = P\alpha$

or $$(\Delta P)1 = \frac{P\alpha}{w}. \qquad **$$

If we premultiply both sides of (4.2.8) by 1^T we have an immediate demonstration of the fact that the weighted mean of the average effects, $1^T P\alpha$, is zero.

The stationary values of the gene frequencies may be found from (4.2.8) by putting $(\Delta P) = 0$, the zero matrix of order k, whence $\hat{P}\hat{\alpha} = 0$, where the circumflex denotes a stationary value. For an internal equilibrium (all $\hat{p}_i > 0$) it follows that $\hat{\alpha}$ must be identically zero, whilst boundary equilibria (one or more $\hat{p}_i = 0$) may be explored by eliminating the corresponding allele from further consideration. For an internal equilibrium, therefore, we have (from (4.2.6))

$$W\hat{P}1 = \hat{w}1, \tag{4.2.9}$$

where $\hat{w}$ is the mean viability at equilibrium, and is simply a normalizing factor. Equation (4.2.9) implies, of course, that every element of the vector $W\hat{P}1$ is equal to $\hat{w}$, and, since W is assumed nonsingular, we have

Theorem 4.2.5. *Internal equilibria*

The internal stationary values of the gene frequencies, if they exist, are given by

$$\hat{P}1 = \hat{w}W^{-1}1, \tag{4.2.10}$$

where (premultiplying by 1^T)
$$1^T \hat{w} W^{-1} 1 = 1$$
so that
$$\hat{w} = (1^T W^{-1} 1)^{-1}. \quad **$$

We note that (4.2.10) implies that there can, at the most, be one internal equilibrium (given that W is non-singular). If W is singular, the equilibria (if any) can be explored directly from (4.2.9).

Let $M_i = \sum_j M_{ij}$, where M_{ij} is the cofactor of w_{ij} in $|W|$, and let $M = \sum_{i,j} M_{ij}$. Then by the ordinary rules for the inversion of a symmetric matrix, (4.2.10) is equivalent to

$$\hat{p}_i = \hat{w} \frac{M_i}{|W|},$$
and
$$\hat{w} = \frac{|W|}{M},$$
so that
$$\hat{p}_i = \frac{M_i}{M}.$$

It follows (Mandel, 1959b, 1970) that an internal equilibrium exists only if the M_i are all non-zero and of the same sign (which will also be the sign of M). We defer consideration of the stability of the equilibrium until the increasing nature of the mean viability has been demonstrated, in the next section.

4.3 PROPERTIES OF THE MEAN VIABILITY

In a manner which parallels the treatment in section 2.3, and whose historical origins are given in section 4.8, we now derive the multi-allelic equivalent of (2.3.12), first given by Li (1969).

Theorem 4.3.1. Li's Theorem

The change in the mean viability from one generation to the next is

$$\Delta w = \frac{v}{w} + 1^T (\Delta P) W (\Delta P) 1.$$

Proof. From (4.2.2)
$$\Delta w = 1^T P' W P' 1 - 1^T P W P 1$$
$$= 1^T [(P + \Delta P) W (P + \Delta P) - P W P] 1$$
$$= 1^T [(\Delta P) W P + P W (\Delta P) + (\Delta P) W (\Delta P)] 1. \quad (4.3.1)$$

PROPERTIES OF THE MEAN VIABILITY

Now from (4.2.8)

$$1^T(\Delta P) = \frac{\alpha^T P}{w} \quad \text{and} \quad (\Delta P)1 = \frac{P\alpha}{w}$$

whence $\quad \Delta w = \dfrac{\alpha^T PWP1}{w} + \dfrac{1^T PWP\alpha}{w} + 1^T(\Delta P)\,W(\Delta P)1.$ (4.3.2)

From (4.2.6) $\qquad PWP1 = P\alpha + wP1$

and thus $\qquad \dfrac{\alpha^T PWP1}{w} = \dfrac{\alpha^T P\alpha}{w} + \alpha^T P1;$

but $\alpha^T P1$ is zero, so that (4.3.2) may now be written

$$\Delta w = 2\frac{\alpha^T P\alpha}{w} + 1^T(\Delta P)\,W(\Delta P)1$$

or, using (4.2.7), $\quad \Delta w = \dfrac{v}{w} + 1^T(\Delta P)\,W(\Delta P)1.$ **

Before leaving Li's Theorem it is interesting to record some alternative forms for the genetic variance. Eliminating α between (4.2.7) and (4.2.8) leads to

$$v = 2w^2 1^T(\Delta P) P^{-1}(\Delta P)1,$$

assuming no p_i zero, which allows Li's Theorem to be written as

$$\Delta w = 2w 1^T(\Delta P) P^{-1}(\Delta P)1 + 1^T(\Delta P)\,W(\Delta P)1$$
$$= 1^T(\Delta P)Z(\Delta P)1,$$

where $\qquad\qquad \{z_{ij}\} = \{w_{ij}\}, \quad i \neq j,$

and $\qquad\qquad \{z_{ii}\} = \{2wp_i^{-1} + w_{ii}\}.$

This formulation might lead to an alternative proof of theorem 4.3.2. The expression for v is interesting because it shows that

$$\frac{v}{2w^2} = \sum_i \frac{(\Delta p_i)^2}{p_i},$$

a fact to which we return in section 4.6, and which shows that the genetic variance is zero at, and only at, equilibrium. If one α in (4.2.7) is replaced by (4.2.6) we obtain

$$v = 2 1^T PWP\alpha,$$

39

since $1^T P \alpha = 0$, and if both αs are replaced we have

$$v = 2(1^T P W P W P 1 - w^2),$$

indicating incidentally the inequality

$$1^T P W P W P 1 \geqslant w^2.$$

We next turn to the multiallelic equivalent of Scheuer and Mandel's Theorem (theorem 2.3.3), that the mean viability increases from one generation to the next. We give some historical comments on this theorem in section 4.8 at the end of the chapter, from which it will be seen that no simple name for the theorem presents itself. We therefore call it:

Theorem 4.3.2. *Increasing mean viability*

The mean viability never decreases from one generation to the next, and remains constant if and only if the gene frequencies are at their equilibrium values.

Proof.

$$w' = 1^T P' W P' 1$$
$$= \frac{1^T P W P W P W P 1}{w^2},$$

from (4.2.4), or

$$w^2 w' = 1^T P W P W P W P 1$$
$$= \tfrac{1}{2}[(1^T P W)(P W P W P)1 + 1^T (P W P W P)(W P 1)]. \tag{4.3.3}$$

Now for non-negative scalar quantities x_i and x_j

$$\tfrac{1}{2}(x_i + x_j) \geqslant \sqrt{(x_i x_j)},$$

so that for non-negative b_{ij}

$$\tfrac{1}{2} \sum_{i,j} b_{ij}(x_i + x_j) \geqslant \sum_{i,j} b_{ij} x_i^{\frac{1}{2}} x_j^{\frac{1}{2}},$$

and this may be written in matrix form

$$\tfrac{1}{2}(x^T B 1 + 1^T B x) \geqslant [x^T]^{\frac{1}{2}} B [x]^{\frac{1}{2}},$$

where $[x]^n$ indicates the vector of elements x_i^n. Applying this

40

inequality to (4.3.3),

$$w^2w' \geqslant [1^TPW]^{\frac{1}{2}}(PWPWP)[WP1]^{\frac{1}{2}}$$
$$= ([1^TPW]^{\frac{1}{2}}PW)P(WP[WP1]^{\frac{1}{2}})$$
$$= 1^TP[WP[WP1]^{\frac{1}{2}}]^2, \qquad (4.3.4)$$

since P is a diagonal matrix.

Furthermore, for proportions p_i, non-negative x_i, and $n \geqslant 1$,

$$\sum_i p_i x_i^n \geqslant (\sum_i p_i x_i)^n,$$

or, in matrix notation,

$$1^TP[x]^n \geqslant (1^TPx)^n. \qquad (4.3.5)$$

Applying (4.3.5) to (4.3.4) with $n = 2$

$$w^2w' \geqslant (1^TPWP[WP1]^{\frac{1}{2}})^2$$
$$= \{(1^TPW)P[WP1]^{\frac{1}{2}}\}^2$$
$$= (1^TP[WP1]^{\frac{3}{2}})^2,$$

again since P is diagonal. Using (4.3.5) again (with $n = \frac{3}{2}$), we finally obtain

$$w^2w' \geqslant \{(1^TPWP1)^{\frac{3}{2}}\}^2$$
$$= (1^TPWP1)^3$$
$$= w^3, \qquad (4.3.6)$$

whence the result $w' \geqslant w$.

For equality it is necessary that, taking the inequality (4.3.5) and assuming $n > 1$ and no p_i zero, all x_i, $i = 1, 2, ..., k$, are equal. In its second application, to obtain (4.3.6), P is the diagonal matrix of gene frequencies and $x = WP1$. Assuming no gene frequency zero (since, if it were, the problem could be reduced by the elimination from consideration of the corresponding gene), for equality every element of $WP1$ must be equal. But, from (4.2.6), $WP1 = \alpha + w1$, so each average effect α_i must be the same, and since $1^TP\alpha = 0$, it must be zero, and we are in gene-frequency equilibrium. This establishes that equilibrium is necessary: that it is sufficient is obvious. **

We may note that this theorem is not dependent on W being non-singular.

It is natural to enquire whether there is a multiallelic equivalent of theorem 2.3.4, that $\Delta w \geqslant v/2w$, and indeed the author conjectured this inequality in 1971. Within a few months, however,

E. A. Thompson had found a triallelic counter-example, the viability matrix being

$$\begin{pmatrix} w_{11} & 0 & 0 \\ 0 & 0 & w_{23} \\ 0 & w_{23} & 0 \end{pmatrix}.$$

Essentially the same counter-example was found by J. F. C. Kingman and published in Senata (1973).

Interestingly enough, for an *unstable* equilibrium we can prove $\Delta w \geqslant v/w$, and we do this in the next section (theorem 4.4.3).

Further comments on the properties of the mean viability are given in chapter 9 on the Fundamental Theorem of Natural Selection.

4.4 STABILITY AND CONVERGENCE

Theorem 4.4.1. *The Equivalence Theorem*

At an internal equilibrium point the gene frequencies identify with the values at the turning point of the mean viability w, and the equilibrium is stable if and only if the turning point is a maximum.

Note. We call an equilibrium 'stable' if *any* small step away from the equilibrium results in a return to it. We here avoid the use of the word 'unstable' because of the possibility of 'saddle-point' situations in which a small step away from the equilibrium in a certain direction may result in a return even though for general perturbations the reverse is true.

Proof. Consider
$$R = w - 2\lambda(\mathbf{1}^{T}P\mathbf{1} - 1),$$

where λ is a Lagrangian multiplier and $\mathbf{1}^{T}P\mathbf{1} = 1$ is the restriction imposed by the p_i summing to unity.

Then
$$\frac{\partial R}{\partial P} = 2WP\mathbf{1} - 2\lambda\mathbf{1} = 0$$

at the turning point of w, which is (4.2.9) with $P = \hat{P}$ and $\lambda = \hat{w}$.

(We should notice that the solution to (4.2.9), given by (4.2.10), might be an external (and therefore genetically meaningless) equilibrium, but that it will still formally correspond to a stationary point of w. That boundary equilibria correspond to stationary points of w regarded as a function on the boundary is clear from considering the problem afresh after eliminating the irrelevant alleles.)

Assuming an internal equilibrium, suppose it is at a maximum of w, and consider a small perturbation from it. Since w must increase in the next generation (by theorem 4.3.2), the new gene-frequency point must lie within the hyper-ellipse of constant w through the old gene-frequency point, and successive points will necessarily be contained in successively smaller nested hyper-ellipses centred on the equilibrium point. Convergence and stability are thus assured; that convergence is global for all internal points is obvious from Li's Theorem (theorem 4.3.1), for as $\Delta P \to 0$ and $\Delta w \to 0$, $v \to 0$, which is only true on approaching the equilibrium point given by the Equivalence Theorem.

Now suppose that the equilibrium is at a saddle-point or a minimum of w. For some or all small perturbations, w will be greater than at the equilibrium point, and no return is possible. By our definition of 'stable' the equilibrium point is not a stable one. **

The condition for the turning point of w to be a maximum was given by Mandel (1959b):

Theorem 4.4.2. *Mandel's Theorem*

A turning point of w is a maximum if and only if

$$(-1)^r \Delta_r < 0, \quad r = 1, 2, ..., k,$$

where Δ_r is the rth principal minor of W.

Proof. Let (δP) be a diagonal matrix of departures from the equilibrium gene frequencies $\hat{P}$. At the point $(\hat{P} + \delta P)$

$$
\begin{aligned}
w &= \mathbf{1}^{\mathrm{T}}(\hat{P} + \delta P) W (\hat{P} + \delta P)\mathbf{1} \\
&= \mathbf{1}^{\mathrm{T}}\hat{P}W\hat{P}\mathbf{1} + \mathbf{1}^{\mathrm{T}}\hat{P}W(\delta P)\mathbf{1} + \mathbf{1}^{\mathrm{T}}(\delta P) W\hat{P}\mathbf{1} \\
&\qquad\qquad + \mathbf{1}^{\mathrm{T}}(\delta P) W(\delta P)\mathbf{1} \\
&= \hat{w} + \mathbf{1}^{\mathrm{T}}(\delta P) W(\delta P)\mathbf{1}
\end{aligned}
$$

since, by (4.2.9), $W\hat{P}\mathbf{1} = \hat{w}\mathbf{1}$ so

$$\mathbf{1}^{\mathrm{T}}(\delta P) W\hat{P}\mathbf{1} = w\mathbf{1}^{\mathrm{T}}(\delta P)\mathbf{1} = 0,$$

and similarly $\mathbf{1}^{\mathrm{T}}\hat{P}W(\delta P)\mathbf{1} = 0.$

w will therefore be at a maximum if and only if the quadratic form $1^T(\delta P) W(\delta P) 1$ is negative for all (δP) satisfying $1^T(\delta P) 1 = 0$, and the condition for this is that W has exactly one positive eigenvalue, from which the theorem follows (Kingman, 1961a, following Mandel, 1959b). **

We may note that since for a stable equilibrium $1^T(\delta P) W(\delta P) 1$ must be negative for all (δP) satisfying $1^T(\delta P) 1 = 0$, it will be negative when $(\delta P) = (\Delta P)$ in particular. But $1^T(\Delta P) W(\Delta P) 1$ is the second term of Li's Theorem (theorem 4.3.1), whence it follows that:

Theorem 4.4.3. Inequalities for the change in mean viability

When there is a stable internal equilibrium

$$\frac{v}{w} \geqslant \Delta w \geqslant 0,$$

but when there is an internal equilibrium which is unstable (in the fullest sense)

$$\Delta w \geqslant \frac{v}{w}.$$

Proof. The second part follows by considering the case where w is at a minimum at equilibrium, which corresponds to complete instability in that the equilibrium is then unapproachable from every direction. **

In view of the fact that we have established the conditions for stability by using the Equivalence Theorem, it is clear that convergence to an internal stable equilibrium is assured from any internal point. When there is no internal stable equilibrium the gene frequency will (except in the rare case of approaching a saddle-point of w along a particular path) change until one gene is lost, and a boundary of the gene-frequency space reached. There may or may not be a stable equilibrium in the reduced space, and the cycle will be repeated until a stable state is reached. Kingman (1961a) asserts that if W has r positive eigenvalues, at least $(r-1)$ alleles must die out before equilibrium is reached.

Sometimes there will be stable equilibria on more than one boundary, but none internally. It is then an unsolved problem to determine the domains of attraction of the several equilibria.

4.5 SPECIAL CASES

We will not pursue the general case of the singular viability matrix, which is treated by Hughes and Seneta (1975), who prove that there is then either no internal equilibrium or uncountably many (all with the same mean viability). However, one singular case may easily be disposed of, that with no dominance in a multiplicative sense

$$(w_{ij}^2 = w_{ii}w_{jj}, \quad i,j = 1 \text{ to } k).$$

For then the viability matrix W may be written

$$W = xx^{\mathrm{T}}, \tag{4.5.1}$$

which is obviously singular, where

$$x_i = \sqrt{w_{ii}}.$$

Hence (4.2.4) becomes

$$P'\mathbf{1} = \frac{Pxx^{\mathrm{T}}P\mathbf{1}}{\mathbf{1}^{\mathrm{T}}Pxx^{\mathrm{T}}P\mathbf{1}}$$

$$= \frac{Px}{\mathbf{1}^{\mathrm{T}}Px}, \tag{4.5.2}$$

since $x^{\mathrm{T}}P\mathbf{1}$ is a scalar quantity in both numerator and denominator.

Thus, just as with two alleles (section 2.2), the gene-frequency change is what it would have been had the gametes, and not the zygotes, differed in viability, in the ratio $\sqrt{w_{11}}:\sqrt{w_{22}}:\ldots:\sqrt{w_{kk}}$ for $a_1:a_2:\ldots:a_k$. In general there will be no internal equilibrium, the population progressing to homozygosity for the gene with the greatest 'viability' from any internal point. Any two genes with the same 'viability' will be indistinguishable by natural selection, and (4.5.2) ensures that their relative proportions will not change. Hence if the greatest viability is shared by more than one gene, there will be uncountably many equilibria corresponding to all possible populations in which only the most viable genes are represented.

Since $w = \mathbf{1}^{\mathrm{T}}Pxx^{\mathrm{T}}P\mathbf{1}$ is a perfect square the $\sqrt{w}$-surface is a flat and the w-surface correspondingly parallel-ruled.

The case of no dominance in an additive sense ($w_{ij} = \frac{1}{2}w_{ii} + \frac{1}{2}w_{jj}$) is in many respects similar, although the viability matrix is not in general singular. Put $x_i = \frac{1}{2}w_{ii}$, whence

$$W = x\mathbf{1}^{\mathrm{T}} + \mathbf{1}x^{\mathrm{T}};$$

45

(4.2.2) then gives

$$w = 1^T P x + x^T P 1 = 2 \, 1^T P x.$$

Thus the w-surface is a flat, and it follows that from any internal point the population will progress to homozygosity for the gene with the greatest 'viability'. The recurrence relation for the gene frequency does not, as in the multiplicative case, ensure that any two genes with the same 'viability' remain in the same relative proportions, but if the greatest viability is shared by more than one gene there will again be uncountably many equilibria, in which only the most viable genes are represented.

4.6 KIMURA'S MAXIMUM PRINCIPLE

In section 4.3 we came across the expression

$$\frac{v}{2w^2} = \sum_i \frac{(\Delta p_i)^2}{p_i} \qquad (4.6.1)$$

relating the genetic variance v to the changes in gene frequency $\{\Delta p_i\}$. In 1958 Kimura introduced his *Maximum Principle in the Genetical Theory of Natural Selection* according to which natural selection causes the gene frequencies to change from $\{p_i\}$ in one generation to $\{p_i + \delta p_i\}$ in the next in such a way that the increase in mean viability (to first order in δp) shall be a maximum under the restriction

$$\sum_i \frac{(\delta p_i)^2}{p_i} = \frac{v}{2w^2}. \qquad (4.6.2)$$

His original treatment was in terms of a continuous-time model, but we shall treat the discrete-generation case. The Principle has found its way into the standard texts, but since no justification for the restriction (4.6.2) has been offered, save that it is true when the actual changes $\{\Delta p_i\}$ are inserted in it (4.6.1), it seems to be of questionable value.

We now show that if the maximization of Δw is to lead to $\delta p_i = \Delta p_i$, all i, then (4.6.2) is, in a sense, a necessary restriction.

To first order in δp the increase in mean viability is

$$\Delta w = 2 \sum_{i,j} (\delta p_i) p_j w_{ij}, \qquad (4.6.3)$$

by analogy with (4.3.1). For any particular Δw this linear function

46

is a flat in the space $\{\delta p_i\}$, and by putting $\delta p_i = \Delta p_i$, all i, we find

$$\Delta w = v/w,$$

the approximate form of Li's Theorem (theorem 4.3.1). Thus

$$2 \sum_{i,j} (\delta p_i) p_j w_{ij} = v/w \qquad (4.6.4)$$

is the flat through the point $\{\Delta p_i\}$. Now if we wish to find a quadratic restriction under which $\delta p_i = \Delta p_i$, all i, is a stationary point of Δw, all we need do is to find a conic to which (4.6.4) is the tangent flat at $\delta p_i = \Delta p_i$. For reasons of symmetry the conic should be an ellipsoid with axes given by the $\{\delta p_i\}$ axes.

This is done by writing (4.6.4) in the form

$$2w \sum_i (\delta p_i)(\Delta p_i)/p_i = v/w, \qquad (4.6.5)$$

to which it is equivalent, by virtue of the expression

$$\Delta p_i = p_i[(1/w) \sum_j p_j w_{ij} - 1], \quad \text{all } i,$$

which is another form of (4.2.4), and then (according to standard theory) replacing each Δp_i by δp_i to obtain

$$\sum_i \frac{(\delta p_i)^2}{p_i} = \frac{v}{2w^2}, \qquad (4.6.2 \; bis)$$

which is (4.6.2), and an ellipsoid since $p_i \geqslant 0$, all i. The argument has not used the fact that the points are restricted to the subspace $\sum_i \delta p_i = 0$, but since $\sum_i \Delta p_i = 0$, this is no problem.

Thus the justification for the restriction (4.6.2) seems to be that it is the only symmetric quadratic function that will do the trick. Since no independent justification has been offered, we conclude that the Maximum Principle simply reflects a feature of the mathematical structure, but adds nothing to an understanding of that structure. The point of interest is (4.6.1) itself. For a further discussion see Edwards (1974) and Narain (1993).

4.7 BAUM AND EAGON'S THEOREM

In section 4.3 we proved that 'the mean viability never decreases from one generation to the next, and remains constant if and only if the gene frequencies are at their equilibrium values' (theorem

4.3.2). An alternative proof of this theorem is provided by the theorem of Baum and Eagon (1967), which is somewhat more general and capable of application to a wider range of genetic models (see section 7.2). We give here, without proof, a restricted version of Baum and Eagon's Theorem, using genetical nomenclature.

Theorem 4.7.1. Baum and Eagon's Theorem

Let $w = \sum_{i,j} p_i p_j w_{ij}$ be a polynomial with non-negative coefficients w_{ij} homogeneous of the second degree in its variables $\{p_i\}$. Let $p = \{p_i\}$ be any point in the gene-frequency space D: $p_i \geq 0$, all i, $\sum_i p_i = 1$. Let p' denote the point of D whose ith coordinate is

$$p_i' = p_i \frac{\partial w}{\partial p_i} \Big/ \sum_i p_i \frac{\partial w}{\partial p_i}, \qquad (4.7.1)$$

where the derivatives are evaluated at p. Then $w' > w$ unless $p' = p$. **

We note that the denominator on the right is simply a normalizing factor, and that since

$$\frac{\partial w}{\partial p_i} = 2 \sum_j p_j w_{ij},$$

$$\sum_i p_i \frac{\partial w}{\partial p_i} = 2 \sum_{i,j} p_i p_j w_{ij} = 2w.$$

Thus we may write (4.7.1) as

$$p_i' = \frac{p_i}{2w} \cdot \frac{\partial w}{\partial p_i}. \qquad (2.4.2\,bis)$$

But this is precisely (2.4.2), which relates the gene frequencies in one generation, $\{p_i\}$, to those in the next, $\{p_i'\}$. w is, of course, simply the mean viability, whence theorem 4.3.2 follows.

4.8   HISTORICAL NOTES

That the mean viability might always increase seems first to have been conjectured in print by Mandel and Hughes (1958), who were influenced by the continuous-time theorem of Fisher (1930), the 'Fundamental Theorem of Natural Selection'. That the conjecture was true was stated by Mandel (1959b) and proved by Scheuer and

Mandel (1959), who gave the explicit expression (2.3.15) for two alleles, thus justifying our naming of theorem 2.3.3.

Simultaneously the conjecture was under consideration by Mulholland and Smith (1959), the proof being obtained by Mulholland, and in the following year a third independent proof was supplied by Atkinson, Watterson and Moran (1960). Subsequently a simplified proof was given by Kingman (1961 b), and it is this, put into matrix notation with the help of J. H. Roger, that we have used in section 4.3. Kingman (1967) considered the inequality further, with special reference to the case of equality. Finally, as noted in section 4.7, Baum and Eagon's (1967) Theorem includes the inequality.

As mentioned in section 2.3 in connection with the case of two alleles, theorem 4.3.2 was first obtained by Li (1969), the two-allele expression (2.3.13) having been given by him in 1967. Our proof of Li's Theorem is essentially a matrix formulation of the proof embedded in Kempthorne and Pollack's (1970) discussion of concepts of fitness, which in turn relies heavily on the approach of Kempthorne (1957) (who did not, however, prove the theorem). The introduction of a matrix notation dates from the 1959 papers listed above, and was first applied fully by Tallis (1966); we have found it convenient to use diagonal matrices for the gametic arrays, following Cannings (1969 b), rather than vectors as in Tallis (1966). It should be noted that there have been several incorrect versions of Li's Theorem published; we do not give the references.

Section 4.4 is due essentially to Mandel (1959 b), but we have quoted theorem 4.4.2 in the form given by Kingman (1961 a), which does not require any new notation. Both Kingman's form and Mandel's equation (30) are saying, in essence, 'if, on crossing out any combination of rows, and the same combination of columns, of W, the determinant of the resulting matrix is positive or negative according as to whether the remaining number of rows (and columns) is odd or even, and if this is true for all combinations, the turning point of w is a maximum'. Thus stated, the theorem has a degree of redundancy, which is indeed why Kingman's and Mandel's forms are equivalent. The interested reader may pursue the matter in Mandel's elegant paper; we shall only do so for the case of three alleles (section 5.3). Mandel (1970) showed that the stability conditions for the discrete-generation model are the same as those found in 1959 (by several authors) for the continuous-time model.

CHAPTER 5

THE SPECIAL CASE OF THREE ALLELES

5.1 GRAPHICAL REPRESENTATION

In chapter 3 we saw how it was advantageous to use homogeneous coordinates to represent the frequencies of the three genotypes that arise at a diallelic locus. It is similarly advantageous to use homogeneous coordinates for the gene frequencies at a triallelic locus, as is often done in order to obtain a graphical representation of how populations differ in respect of such frequencies (initially by Streng, 1926). Although homogeneous coordinates may theoretically be used for any number of alleles, with just three it is easy to depict the mean viability function using contours, and hence to give a geometrical interpretation of the conditions for the existence and stability of equilibria found by Mandel (1959b). The treatment of this chapter amounts, in part, to a reworking of Feller (1969) using homogeneous coordinates. Section 3.1 should be consulted for some remarks on such coordinates.

5.2 THE SHAPE OF THE MEAN VIABILITY SURFACE

For three alleles the mean viability is, from (2.3.1),

$$w = w_{11} p_1^2 + w_{22} p_2^2 + w_{33} p_3^2 + 2w_{23} p_2 p_3 \\ + 2w_{13} p_1 p_3 + 2w_{12} p_1 p_2, \qquad (5.2.1)$$

which, since $(p_1 + p_2 + p_3)^2 = 1$, may be thrown into homogeneous form as

$$(w_{11} - w)p_1^2 + (w_{22} - w)p_2^2 + (w_{33} - w)p_3^2 \\ + 2(w_{23} - w)p_2 p_3 + 2(w_{13} - w)p_1 p_3 \\ + 2(w_{12} - w)p_1 p_2 = 0. \qquad (5.2.2)$$

For any particular value of w from among the set of possible values (that is, those values generated by all p_1, p_2, p_3 such that

$p_1 + p_2 + p_3 = 1$, negative values allowed) this equation represents a conic.† In what follows w is to be interpreted as any one of the possible values other than the maximum or minimum of (5.2.1) (if such exist, and unless otherwise indicated).

By a standard result, the centre (x, y, z) is the pole of the line at infinity $p_1 + p_2 + p_3 = 0$, which is

$$(w_{11} - w) x + (w_{12} - w) y + (w_{13} - w) z$$
$$= (w_{12} - w) x + (w_{22} - w) y + (w_{23} - w) z$$
$$= (w_{13} - w) x + (w_{23} - w) y + (w_{33} - w) z$$

or, adding $wx + wy + wz$ (the constant w) to all sides,

$$w_{11} x + w_{12} y + w_{13} z = w_{12} x + w_{22} y + w_{23} z$$
$$= w_{13} x + w_{23} y + w_{33} z. \tag{5.2.3}$$

Since this does not involve w, (5.2.2) represents a concentric family of conics. It is readily shown that (5.2.3) is another form of (4.2.10), and has the same solution; each part of (5.2.3) equals $\hat{w}$. A demonstration of this follows in section 5.3.

If in (5.2.1) we put

$$\left.\begin{array}{l} \sigma_{23} = 2w_{23} - w_{22} - w_{33} \\ \sigma_{13} = 2w_{13} - w_{11} - w_{33} \\ \sigma_{12} = 2w_{12} - w_{11} - w_{22} \end{array}\right\}, \tag{5.2.4}$$

the σ_{ij} thus being measures of pairwise dominance in an additive sense,

$$w = \sigma_{23} p_2 p_3 + \sigma_{13} p_1 p_3 + \sigma_{12} p_1 p_2$$
$$+ w_{11} p_1 + w_{22} p_2 + w_{33} p_3. \tag{5.2.5}$$

If all the σ_{ij} are zero (5.2.5) is a straight line and, representing w by the dimension perpendicular to the plane of the reference triangle, the w-surface is a plane, inclined in general (see section 4.5). We defer the cases of one or two σ_{ij} zero until later.

Next, in order to show that each conic (5.2.5) is of the same shape and orientation, or *homothetic*, we use the proposition that if any two conics can be expressed in inhomogeneous form with identical quadratic terms, they are homothetic. This is easily seen to be the

† It will already be obvious to readers familiar with homogeneous coordinates that, for varying w, (5.2.2) represents a family of concentric similar conics, but we shall show this in stages.

case, for if in each we put $p_3 = 1 - p_1 - p_2$ the two conics still have identical quadratic terms, and on a particular choice of reference triangle p_1 and p_2 are simply Cartesian coordinates, in terms of which the proposition is known to be true. Other reference triangles induce linear transformations which will affect the two conics' shape and orientation equally.

It follows that the conic

$$\sigma_{23} p_2 p_3 + \sigma_{13} p_1 p_3 + \sigma_{12} p_1 p_2 = 0 \qquad (5.2.6)$$

has the same shape and orientation as (5.2.5), but it is more convenient to work with, since it clearly passes through the three vertices of the reference triangle. Since all the conics (5.2.2) formed by varying w thus have the same shape, orientation, and (as we have seen) centre, the contours of the mean viability surface are a concentric family of homothetic conics.

Theorem 5.2.1. *Contours of the mean viability surface*

The contours of the mean viability surface are a concentric family of homothetic conics, being ellipses, parabolae or hyperbolae according as to whether

$$\sigma_{23}^2 + \sigma_{13}^2 + \sigma_{12}^2 - 2\sigma_{13}\sigma_{12} - 2\sigma_{23}\sigma_{12} - 2\sigma_{23}\sigma_{13}$$

is less than, equal to, or greater than, zero.

Proof. That they are concentric conics homothetic to (5.2.6) has already been shown. To find when (5.2.6) is an ellipse, etc., we consider its intercept with the line at infinity, $p_1 + p_2 + p_3 = 0$. It is a standard result that the line at infinity intersects the conic

$$a p_1^2 + b p_2^2 + c p_3^2 + 2f p_2 p_3 + 2g p_1 p_3 + 2h p_1 p_2 = 0$$

at no, one, or two points according as to whether

$$\begin{vmatrix} a & h & g & 1 \\ h & b & f & 1 \\ g & f & c & 1 \\ 1 & 1 & 1 & 0 \end{vmatrix}$$

is less than, equal to, or greater than, zero, conditions which indicate an ellipse, a parabola, and a hyperbola, respectively. Applying this to (5.2.6) we find

$$\begin{vmatrix} 0 & \sigma_{12} & \sigma_{13} & 1 \\ \sigma_{12} & 0 & \sigma_{23} & 1 \\ \sigma_{13} & \sigma_{23} & 0 & 1 \\ 1 & 1 & 1 & 0 \end{vmatrix}$$

$$= \sigma_{23}^2 + \sigma_{13}^2 + \sigma_{12}^2 - 2\sigma_{13}\sigma_{12} - 2\sigma_{23}\sigma_{12} - 2\sigma_{23}\sigma_{13},$$

and the theorem is proved. **

An alternative proof is to apply the above standard result directly to (5.2.2), obtaining

$$D = \begin{vmatrix} w_{11} & w_{12} & w_{13} & 1 \\ w_{12} & w_{22} & w_{23} & 1 \\ w_{13} & w_{23} & w_{33} & 1 \\ 1 & 1 & 1 & 0 \end{vmatrix}, \tag{5.2.7}$$

in which w has been eliminated by adding to each of the first three columns the fourth column multiplied by w. Substituting for the w_{ij} $(i \neq j)$ in terms of the σ_{ij} and w_{ii} by (5.2.4), the w_{ii} and w_{ij} are all eliminated, leading to

$$4D = \sigma_{23}^2 + \sigma_{13}^2 + \sigma_{12}^2 - 2\sigma_{13}\sigma_{12} - 2\sigma_{23}\sigma_{12} - 2\sigma_{23}\sigma_{13}.$$

It is natural that the condition should be expressible in terms of the σ_{ij} alone, since adding a constant to each viability changes only w in (5.2.2), and does not change the σ_{ij}.

Another form for D is in terms of quantities that have already been defined in section 4.2:

$$\left. \begin{aligned} M_{11} &= w_{22}w_{33} - w_{23}^2 \\ M_{12} &= w_{13}w_{23} - w_{12}w_{33} \\ M_{13} &= w_{12}w_{23} - w_{22}w_{13} \\ M_{22} &= w_{11}w_{33} - w_{13}^2 \\ M_{23} &= w_{12}w_{13} - w_{11}w_{23} \\ M_{33} &= w_{11}w_{22} - w_{12}^2 \end{aligned} \right\} , \tag{5.2.8}$$

the cofactors in $|W|$. Straightforward algebra leads to

$$-D = M_{11} + M_{22} + M_{33} + 2M_{23} + 2M_{13} + 2M_{12}$$
$$= M_1 + M_2 + M_3 = M, \tag{5.2.9}$$

where $\qquad M_i = \sum_j M_{ij} \quad$ and $\quad M = \sum_{i,j} M_{ij}.$

If any two σ_{ij} are zero, say σ_{23} and σ_{13}, (5.2.6) becomes $\sigma_{12}p_1p_2 = 0$ which, by theorem 5.2.1, is a hyperbola. Indeed it is the degenerate hyperbola consisting of the lines $p_1 = 0$ and $p_2 = 0$; thus for varying w (5.2.5) must then be a family of hyperbolae with asymptotes parallel to $p_1 = 0$ and $p_2 = 0$. If only one σ_{ij} is zero, say σ_{12}, (5.2.6) becomes

$$\sigma_{23}p_2p_3 + \sigma_{13}p_1p_3 = 0$$

which, again by theorem 5.2.1, is a hyperbola, this time the degenerate hyperbola

$$p_3 = 0 \quad \text{and} \quad \sigma_{23}p_2 + \sigma_{13}p_1 = 0.$$

Hence for varying w (5.2.5) must then be a family of hyperbolae with asymptotes parallel to these two lines.

The case of all σ_{ij} zero has already been mentioned, but it may also be noted that then $|W| = 0$, and W is singular.

We now prove

Theorem 5.2.2. Feller's Theorem, part I

The contours are ellipses if and only if the σ_{ij} are of the same sign and there exists a triangle with sides L_i such that

$$L_1^2 : L_2^2 : L_3^2 = \sigma_{23} : \sigma_{13} : \sigma_{12}.$$

Proof. By theorem 5.2.1 the contours are ellipses if and only if

$$\sigma_{23}^2 + \sigma_{13}^2 + \sigma_{12}^2 - 2\sigma_{13}\sigma_{12} - 2\sigma_{23}\sigma_{12} - 2\sigma_{23}\sigma_{13} < 0. \quad (5.2.10)$$

If one or two σ_{ij} are zero (5.2.10) cannot hold, as we have seen, and neither can there be a non-degenerate triangle. If one is negative, say σ_{12}, and the others positive, (5.2.10) is

$$\sigma_{23}^2 + \sigma_{13}^2 + \rho_{12}^2 + 2\sigma_{13}\rho_{12} + 2\sigma_{23}\rho_{12} - 2\sigma_{23}\sigma_{13} < 0,$$

where $\rho_{12} = -\sigma_{12}$ and all the values are now positive. It follows that

$$(\sigma_{23} + \sigma_{13} + \rho_{12})^2 < 4\sigma_{23}\sigma_{13},$$

whence $\qquad\qquad \sigma_{23} + \sigma_{13} + \rho_{12} < 2\sqrt{(\sigma_{23}\sigma_{13})},$

the positive root being taken, and

$$(\sqrt{\sigma_{23}} - \sqrt{\sigma_{13}})^2 < -\rho_{12},$$

an impossibility. If just two σ_{ij} are negative the argument can be repeated, leading to a similar impossibility. Thus a necessary condition for an ellipse is that all the σ_{ij} be the same sign. Without loss of generality, we assume them all positive.

Now the left side of (5.2.10) may be factored into

$$(\sqrt{\sigma_{23}} - \sqrt{\sigma_{13}} - \sqrt{\sigma_{12}})(\sqrt{\sigma_{13}} - \sqrt{\sigma_{23}} - \sqrt{\sigma_{12}})$$
$$\times (\sqrt{\sigma_{12}} - \sqrt{\sigma_{23}} - \sqrt{\sigma_{13}})(\sqrt{\sigma_{23}} + \sqrt{\sigma_{13}} + \sqrt{\sigma_{12}}), \quad (5.2.11)$$

positive roots being taken. For this to be negative, either all the first three factors must be negative, or just one of them.

Suppose only the first negative. Then

$$\sqrt{\sigma_{13}} > \sqrt{\sigma_{23}} + \sqrt{\sigma_{12}}$$

and

$$\sqrt{\sigma_{12}} > \sqrt{\sigma_{23}} + \sqrt{\sigma_{13}},$$

a joint impossibility. Therefore all three would have to be negative:

$$\left. \begin{array}{l} \sqrt{\sigma_{23}} < \sqrt{\sigma_{13}} + \sqrt{\sigma_{12}} \\ \sqrt{\sigma_{13}} < \sqrt{\sigma_{23}} + \sqrt{\sigma_{12}} \\ \sqrt{\sigma_{12}} < \sqrt{\sigma_{23}} + \sqrt{\sigma_{13}} \end{array} \right\}. \quad (5.2.12)$$

But these are simply the triangle inequalities for sides L_1, L_2 and L_3, and Feller's Theorem, part I, is proved. **

Theorem 5.2.2 is a slightly more graphic, but also more complex, statement of theorem 5.2.1.

By virtue of the factorization (5.2.11), and assuming that not all the σ_{ij} are zero (a case already dealt with), the contours will be parabolae when, for example,

$$\sqrt{\sigma_{23}} = \sqrt{\sigma_{13}} + \sqrt{\sigma_{12}}.$$

If the contours are ellipses we can ask with respect to what reference triangle they would become circles. In contemplating a reference triangle which is not equilateral we must decide between the use of trilinear coordinates (the lengths of the perpendiculars from the given point to the sides of the reference triangle) and the use of areal coordinates (the areas of the triangles formed by the lines from the given point to the vertices of the reference triangles).† We choose the latter because areal coordinates are unaffected by

† See section 3.1.

orthogonal projection, which we shall use to project the ellipse into a circle and establish:

Theorem 5.2.3. Feller's Theorem, part II

Elliptical contours may be rendered circular if, using areal coordinates, the reference triangle is chosen with sides L_1, L_2 and L_3.

Proof. By theorem 5.2.2 we note that such a reference triangle exists. A known result in areal coordinates is that if the reference triangle has sides a, b and c the circumscribing circle is

$$a^2 p_2 p_3 + b^2 p_1 p_3 + c^2 p_1 p_2 = 0. \qquad (5.2.13)$$

Identifying (5.2.6) and (5.2.13) we see that

$$\left. \begin{array}{l} a^2 = \sigma_{23} \\ b^2 = \sigma_{13} \\ c^2 = \sigma_{12} \end{array} \right\}$$

ensures the required result, for the ellipse (5.2.6) circumscribing the reference triangle with equal sides is the same shape and orientation as the elliptical contours, and the orthogonal projection implied by the new reference triangle takes this ellipse into a circle and the equilateral triangle into an inscribed triangle with sides L_1, L_2 and L_3. **

If the contours are parabolae then the reference triangle is degenerate; as is well known, no orthogonal projection will take parabolae into circles. If the contours are hyperbolae, then by theorem 5.2.2 a reference triangle with the necessary sides does not exist. We can, however, always render a hyperbola rectangular by orthogonal projection, but we shall not pursue this, first because an element of arbitrariness enters into the choice of the projection (and hence the reference triangle) and secondly because our primary interest is in cases of stable equilibrium, which have elliptical contours. There is the further factor that, notwithstanding the availability of these interesting projections, we prefer to continue to use an equilateral triangle of reference for graphical presentation. Feller (1969) examines the hyperbolic cases in more detail.

Additionally we may note that a transformation to new viabilities w'_{ij}, where

$$w_{ij} = \lambda w'_{ij} + \mu, \quad w'_{ij} \geqslant 0,$$

leads to the same geometrical figure with the contour lines re-labelled, since substitution in (5.2.1) gives

$$w = \lambda w' + \mu.$$

Theorem 5.2.4. Sections of the mean viability surface

Along any line the graph of the mean viability w (5.2.1) is a parabola.

Proof. Solving for p_2 and p_3 in terms of p_1 between the equation for the line and $p_1 + p_2 + p_3 = 1$, and substituting the results in (5.2.1), (5.2.1) becomes a quadratic in p_1, which represents a parabola. **

Theorem 5.2.5. Curvature of the mean viability surface in the elliptical case

With elliptical contours, if the σ_{ij} are positive the mean viability surface is concave,† and if the σ_{ij} are negative it is convex.

Proof. By theorem 5.2.4 every section of the surface is a parabola and, since the surface has elliptical contours, if any section is concave so is the whole surface. Consider $\sigma_{12} > 0$, which implies

$$2w_{12} > w_{11} + w_{22}$$

and hence $\tfrac{1}{4}w_{11} + \tfrac{1}{2}w_{12} + \tfrac{1}{4}w_{22} > \tfrac{1}{2}(w_{11} + w_{22}).$

The left side is the mean viability at $(\tfrac{1}{2}, \tfrac{1}{2}, 0)$, which thereby exceeds the mean of the mean viabilities at $(1, 0, 0)$ and $(0, 1, 0)$. Thus along $p_3 = 0$ the surface is concave, and hence everywhere. The reverse holds if $\sigma_{12} < 0$. ** For another proof see Edwards (1977).

We may note in passing that the stationary point on $p_1 = 0$ is at $p_2 = (w_{23} - w_{33})/\sigma_{23}$, and similarly for the other boundaries $p_2 = 0$ and $p_3 = 0$, whence, using Feller's Theorem, part II, the circular contours in the transformed space may be located geometrically, the centre of the family of circles being at the join of the perpendiculars to the three sides of the reference triangle at these stationary points.

5.3   CONDITIONS FOR STABLE EQUILIBRIUM

From the Equivalence Theorem (theorem 4.4.1) we know that at an internal equilibrium point the gene frequencies identify with the

† That is, concave when viewed from below.

values at the turning point of w, and that the equilibrium is stable if and only if the turning point is a maximum.

It immediately follows that for an internal stable equilibrium to exist it is necessary and sufficient that

(1) the centre (5.2.3) of the family of conics be within the reference triangle;

(2) the family of conics be ellipses; and

(3) w be a maximum and not a minimum at the centre.

Taking these points in order, we first require that x, y and z in

$$w_{11}x + w_{12}y + w_{13}z = w_{12}x + w_{22}y + w_{23}z$$
$$= w_{13}x + w_{23}y + w_{33}z \quad (5.2.3 \ bis)$$

be all positive. Now from the definition of the average effects

$$\alpha_i = w_{1i}p_1 + w_{2i}p_2 + w_{3i}p_3 - w, \quad (2.3.4 \ bis)$$

so the centre is where the average effects are all equal. But each is then zero (section 4.2), whence at the centre

$$\left.\begin{array}{l} w_{11}x + w_{12}y + w_{13}z = \hat{w} \\ w_{12}x + w_{22}y + w_{23}z = \hat{w} \\ w_{13}x + w_{23}y + w_{33}z = \hat{w} \end{array}\right\},$$

whose solution is already known (section 4.2). The condition that x, y and z be all positive is thus that M_1, M_2 and M_3 must all have the same sign.

The second point, that the conics be ellipses, is met if and only if

$$\sigma_{23}^2 + \sigma_{13}^2 + \sigma_{12}^2 - 2\sigma_{13}\sigma_{12} - 2\sigma_{23}\sigma_{12} - 2\sigma_{23}\sigma_{13}$$
$$= -4(M_1 + M_2 + M_3) < 0, \quad (5.3.1)$$

by theorem 5.2.1 and the subsequent arguments, and the third point, that the mean viability be a maximum rather than a minimum at the centre, is met if and only if the σ_{ij} are all positive (theorem 5.2.5).

Thus we have the following theorem:

Theorem 5.3.1. *Conditions for stable equilibrium*

The necessary and sufficient conditions for an internal stable equilibrium are that each of M_1, M_2, M_3, σ_{23}, σ_{13} and σ_{12} be positive. **

In view of (5.3.1) these six quantities represent only five independent parameters, in spite of being functions of the six independent viabilities w_{ij}. This is because both the σ_{ij} and the M_i are insensitive to the addition of a constant to all the w_{ij}, so that the smallest of the w_{ij} may be assigned the value zero without loss of generality.

It remains to show that these conditions are equivalent to those of Owen (1954), Mandel (1959b) and Kingman (1961a) (see our theorem 4.4.2). Owen's and Kingman's results presume the existence of an internal equilibrium, so to theirs we must add:

$$M_1, \ M_2 \ \text{and} \ M_3 \ \text{must all have the same sign.} \qquad (5.3.2)$$

Kingman's form for three alleles is, in our notation,

$$M_{33} < 0, \quad \Delta = |W| > 0,$$

which, as Mandel notes (and symmetry demands), is equivalent to Owen's
$$M_{11}, \ M_{22}, \ M_{33} < 0, \quad \Delta > 0, \quad \text{where} \ \Delta = |W|. \qquad (5.3.3)$$
Mandel's other form for stability (without reference to whether the equilibrium is internal) is

$$\begin{vmatrix} w_{11} & w_{13} & 1 \\ w_{13} & w_{33} & 1 \\ 1 & 1 & 0 \end{vmatrix} > 0, \quad D < 0,$$

or $\sigma_{13} > 0, D < 0$ in our notation. This, as he notes, is equivalent to

$$\sigma_{23}, \sigma_{13}, \sigma_{12} > 0, \quad D < 0, \qquad (5.3.4)$$

and is simply our theorems 5.2.1 and 5.2.5, which, together with (5.3.2), lead to theorem 5.3.1 in view of (5.2.9). Similarly theorem 5.3.1 implies (5.3.2) and (5.3.4).

We conclude by proving that (5.3.4) implies (5.3.3) unconditionally, but (5.3.3) implies (5.3.4) subject to (5.3.2). Consider

$$D = \begin{vmatrix} w_{11} & w_{12} & w_{13} & 1 \\ w_{12} & w_{22} & w_{23} & 1 \\ w_{13} & w_{23} & w_{33} & 1 \\ 1 & 1 & 1 & 0 \end{vmatrix} \qquad (5.2.7 \ bis)$$

and its adjugate, in which we will only enter the elements needed for the present proof:

$$\text{adj} \ D = \begin{vmatrix} \cdot & \cdot & \cdot & \cdot \\ \cdot & \cdot & \cdot & \cdot \\ \cdot & \cdot & \sigma_{12} & -M_3 \\ \cdot & \cdot & -M_3 & \Delta \end{vmatrix}. \qquad (5.3.5)$$

3-2

That the off-diagonal element given is

$$
-\begin{vmatrix} w_{11} & w_{12} & w_{13} \\ w_{12} & w_{22} & w_{23} \\ 1 & 1 & 1 \end{vmatrix} = -M_3
$$

is convenient, though not actually essential to the proof, which only requires that its square be positive.

Now Jacobi's Theorem is that any minor of order k in adj D is equal to the complementary cofactor in D, multiplied by D^{k-1}. Applying this to the minor of order 2 given in (5.3.5),

$$\sigma_{12}\Delta - M_3^2 = (w_{11}w_{22} - w_{12}^2)\,D$$

or $$\sigma_{12}\Delta - M_3^2 = M_{33}D. \qquad (5.3.6)$$

Consider (5.3.4); $\sigma_{13} > 0$ is

$$w_{13} > \tfrac{1}{2}(w_{11} + w_{33})$$

which implies $$w_{13} > \sqrt{(w_{11}w_{33})},$$

by the inequality of the arithmetic and geometric means, and is $M_{22} < 0$. Thus σ_{23}, σ_{13}, $\sigma_{12} > 0$ imply M_{11}, M_{22}, $M_{33} < 0$. Under these conditions (5.3.6) ensures that if D is negative Δ must be positive, completing the proof that (5.3.4) implies (5.3.3).

On the other hand, starting with (5.3.3) we cannot arrive at (5.3.4) without using (5.3.2): an expression for Δ is

$$\Delta = w_{11}M_1 + w_{12}M_2 + w_{13}M_3,$$

which shows that if M_1, M_2 and M_3 have the same sign it is that of Δ, and hence opposite to that of

$$D = -(M_1 + M_2 + M_3). \qquad (5.2.9\ bis)$$

Thus Δ positive implies D negative, whereupon (5.3.6) ensures that σ_{12} is positive. The inequality (5.3.4) is completed by repeating the argument with interchanges of indices.

5.4 EXAMPLES OF POSSIBLE SYSTEMS

It is possible to classify the systems that can give rise to stable equilibria in a number of different ways. Thus Mandel (1959b) considers the order relations amongst the viabilities and concludes 'that there are only four essentially different systems', whilst Feller (1969) prefers a geometrical enumeration. The major differences are deter-

Type 1

Type 2

Type 3

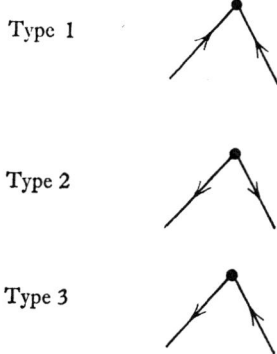

Figure 5.1. Types of corners of the reference triangle; arrows indicate directions of increasing mean viability.

mined by the order relations between each heterozygote and its associated homozygotes, which provide a classification fine enough for the present purpose of creating examples.

We wish to enumerate, for stable equilibria only, the possible relationships that the triangle of reference might have to the circular contours of mean viability, following the transformation given by part II of Feller's Theorem (theorem 5.2.3). Indicating by an arrow a direction in which the mean viability increases, there can only be three essentially different kinds of corner for the triangle of reference, corresponding to the homozygote's viability being greater than both, one, or neither of the associated heterozygotes' (figure 5.1).

If any corner is of Type 1 an internal stable equilibrium cannot occur, for the maximum of the mean viability surface must lie beyond the lines through the corner perpendicular to each side (figure 5.2). Similarly, if any corner of Type 3 has an acute angle, an internal stable equilibrium cannot occur (figure 5.3).

There are thus only two basic forms admitting internal stable equilibria (figure 5.4). In the first, each heterozygote has a greater viability than its associated homozygotes, or, what is the same thing, each homozygote has a lesser viability than its associated heterozygotes. (Figure 5.4 (i) is drawn with all the angles of the triangle acute; a slightly different figure results if one of the angles is obtuse.) Note that these order relations amongst the viabilities do not ensure the existence of a stable equilibrium, for the maximum of the mean viability surface may still be outside the triangle.

61

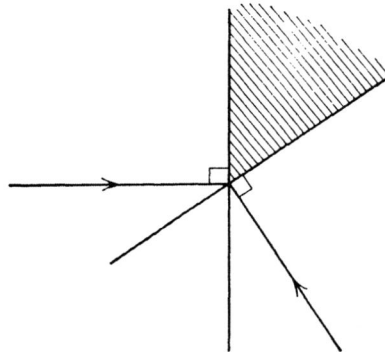

Figure 5.2. The shaded area indicates where the maximum must lie for a corner of Type 1, and that it is necessarily outside the triangle.

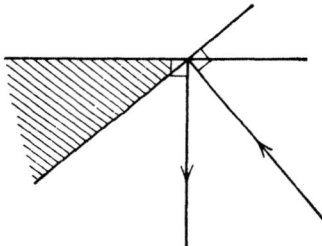

Figure 5.3. The shaded area indicates where the maximum must lie for an acute-angled corner of Type 3, and that it is necessarily outside the triangle.

In the second form, two of the homozygotes have lesser viabilities than their associated heterozygotes, but the third is intermediate, and the corresponding angle obtuse. Thus, if a_3a_3 is this third homozygote,

$$w_{11} < w_{12}, w_{13}$$

$$w_{22} < w_{12}, w_{23}$$

$$w_{13} < w_{33} < w_{23} \quad \text{(say)},$$

and
$$\sigma_{12} > \sigma_{13} + \sigma_{23},$$

since, by Feller's Theorem, part II (theorem 5.2.3), the triangle has sides $\sqrt{\sigma_{23}}$, $\sqrt{\sigma_{13}}$, $\sqrt{\sigma_{12}}$, the last-named side being opposite the obtuse

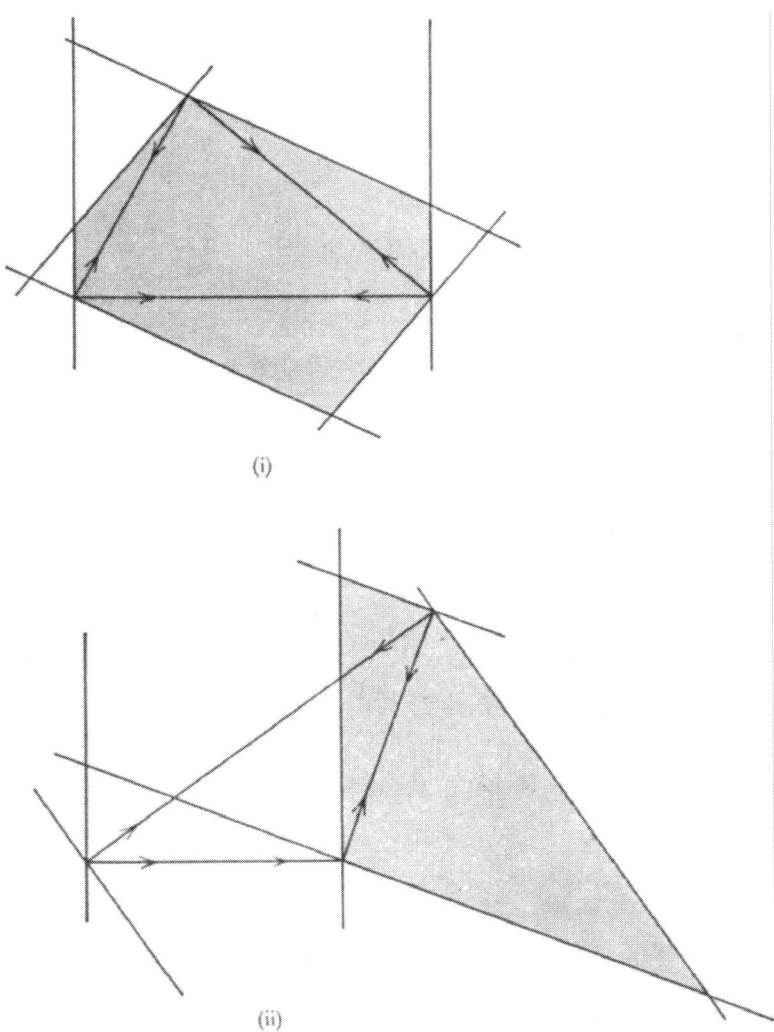

(i)

(ii)

Figure 5.4. Two basic types of reference triangle admitting an internal stable equilibrium. With respect to circular contours, the equilibrium must lie in the shaded area.

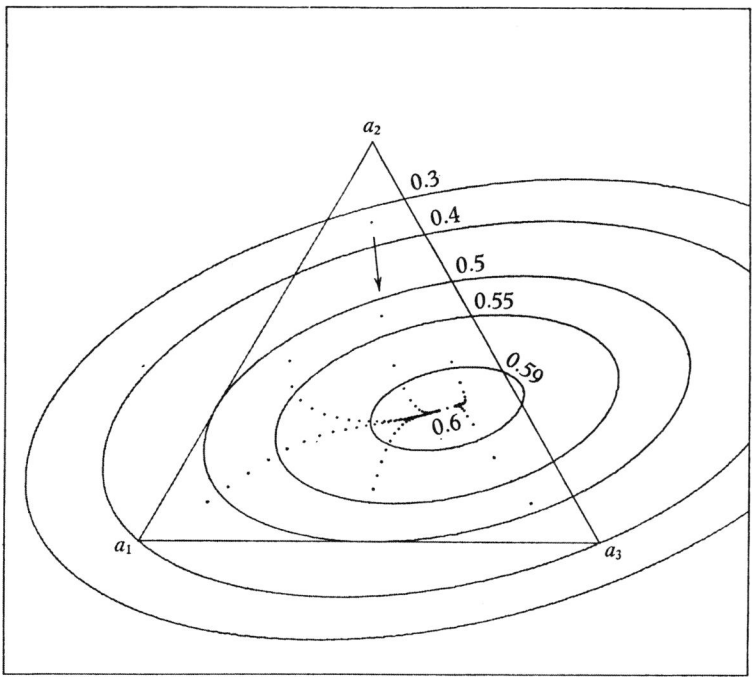

Figure 5.5. Mean viability function and population trajectories
(see table 5.1).

angle. The latter condition is equivalent to

$$w_{12} + w_{33} > w_{13} + w_{23}.$$

Thus the higher viability of each heterozygote than its associated homozygotes is neither necessary nor sufficient for an internal stable equilibrium (Owen, 1954). Mandel's (1959b) classification, which is finer than the present one since it considers all the order relations, subdivides our first form according as to whether or not any one of the heterozygotes is less viable than the non-associated homozygote, and subdivides our second form according as to whether the 'anomalous' heterozygote is or is not less viable than the non-associated homozygote.

Figure 5.5 shows an example of a stable equilibrium of the first kind, and figure 5.6 of the second kind. Figure 5.7 shows an example of the first kind in which there is no internal stable equilibrium. Figure 5.8 shows an example of hyperbolic contours. Each figure gives the contours of the mean viability surface (note that they are

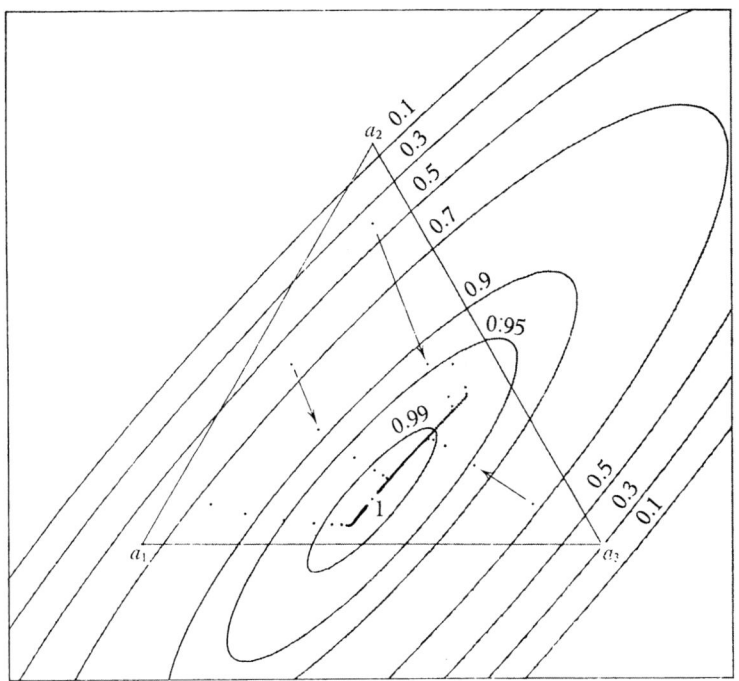

Figure 5.6. Mean viability function and population trajectories
(see table 5.1).

Table 5.1. *Viability matrices and equilibrium points for the figures
in section 5.4*

Figures 5.5 and 5.12
$$\begin{pmatrix} 0.4 & 0.7 & 0.6 \\ 0.7 & 0.1 & 0.9 \\ 0.6 & 0.9 & 0.4 \end{pmatrix}$$
Equilibrium point ($\frac{1}{6}$, $\frac{1}{3}$, $\frac{1}{2}$); stable.

Figures 5.6 and 5.13
$$\begin{pmatrix} 0.625 & 0.5 & 1.5 \\ 0.5 & 0 & 1.75 \\ 1.5 & 1.75 & 0.3125 \end{pmatrix}$$
Equilibrium point ($\frac{4}{9}$, $\frac{1}{9}$, $\frac{4}{9}$); stable.

Figures 5.7 and 5.14
$$\begin{pmatrix} 0 & 10 & 5 \\ 10 & 0 & 4 \\ 5 & 4 & 0 \end{pmatrix}$$
No internal equilibrium point; external equilibrium at ($\frac{44}{70}$, $\frac{45}{70}$, $-\frac{19}{70}$); effective
stable equilibrium at ($\frac{1}{2}$, $\frac{1}{2}$, 0).

Figures 5.8 and 5.15
$$\begin{pmatrix} 2 & 2 & 1 \\ 2 & 1 & 2 \\ 1 & 2 & 2 \end{pmatrix}$$
Equilibrium point ($\frac{1}{3}$, $\frac{1}{3}$, $\frac{1}{3}$); unstable (saddle-point); effective stable equilibria
at (0, 0, 1) and (1, 0, 0).

65

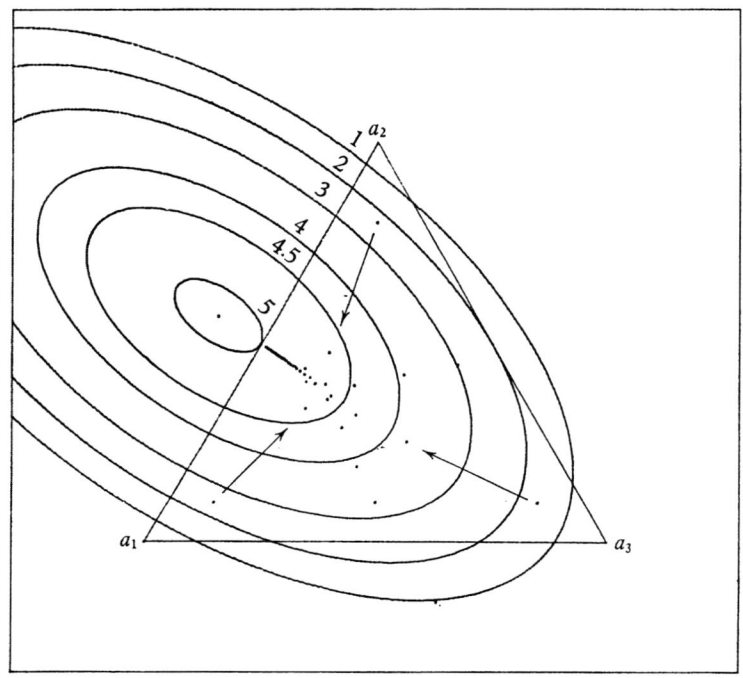

Figure 5.7. Mean viability function and population trajectories
(see table 5.1).

not at equal height intervals) and the trajectories of populations
starting from selected initial positions. It is interesting to see how
often a gene frequency does not change monotonically, perhaps
increasing and then decreasing, and how very 'attractive' a line close
to the major axis of an eccentric ellipse is. Indeed, points close to
such an axis are almost in a state of neutral stability, and we may
infer that should a number of independent populations be subject
to the same viability matrix, a triangular-chart plot of their gene frequencies would reveal a clustering along the major axis, from which
some features of the viability matrix might be estimated.

The examples derive from Mulholland and Smith (1959) [figure
5.5], Owen (1959) [figures 5.6 and 5.7] and Blakley (1967) [figure
5.8], and the viability matrices are given in table 5.1.

In figures 5.9 and 5.10 we give examples of how the mean viability
may change quite erratically, even though always increasing. Figure
5.9 corresponds to the population which started near the apex of

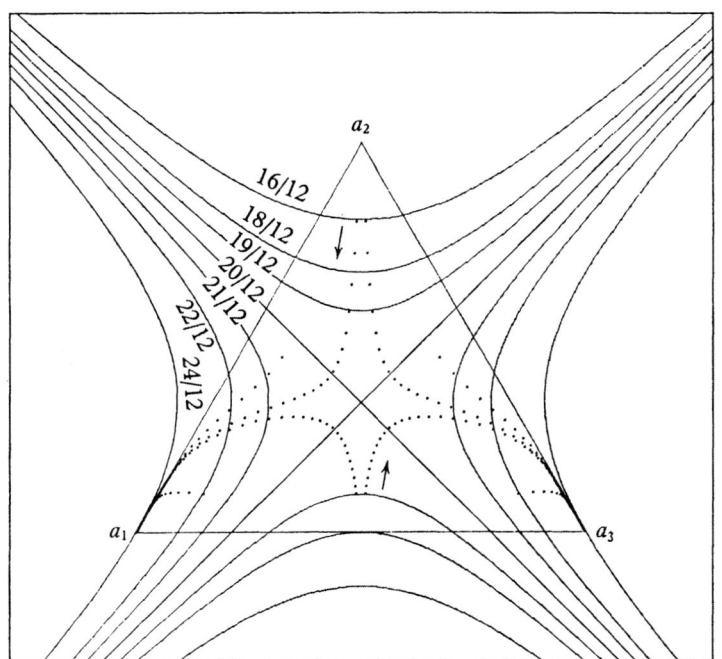

Figure 5.8. Mean viability function and population trajectories (see table 5.1).

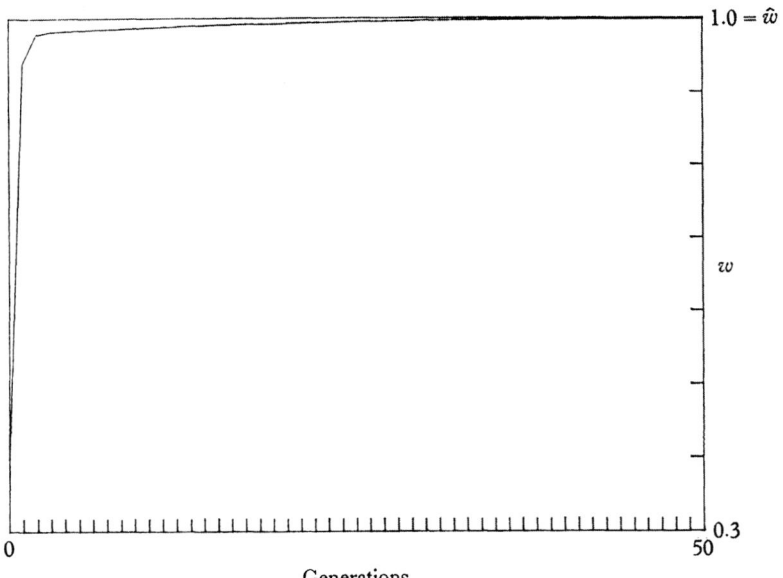

Figure 5.9. Successive values of the mean viability for a population starting at (0.1, 0.8, 0.1) in figure 5.6.

67

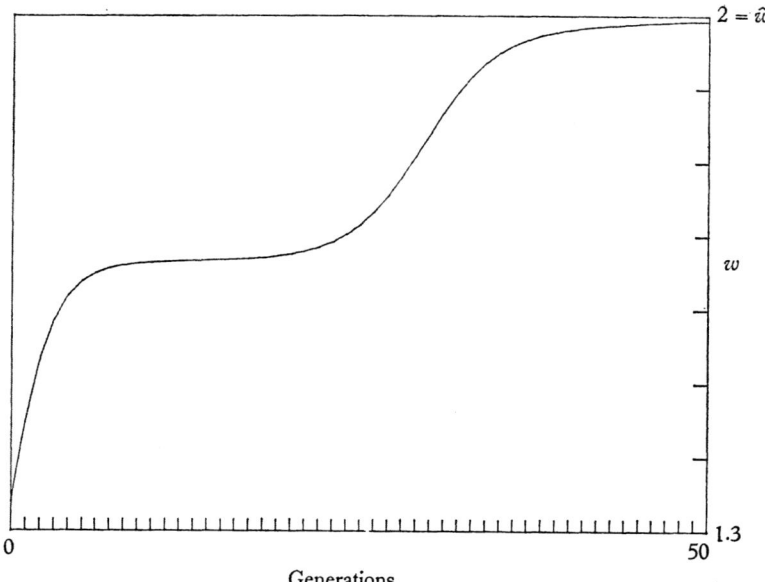

Figure 5.10. Successive values of the mean viability for a population starting at (0.099, 0.8, 0.101) in figure 5.8.

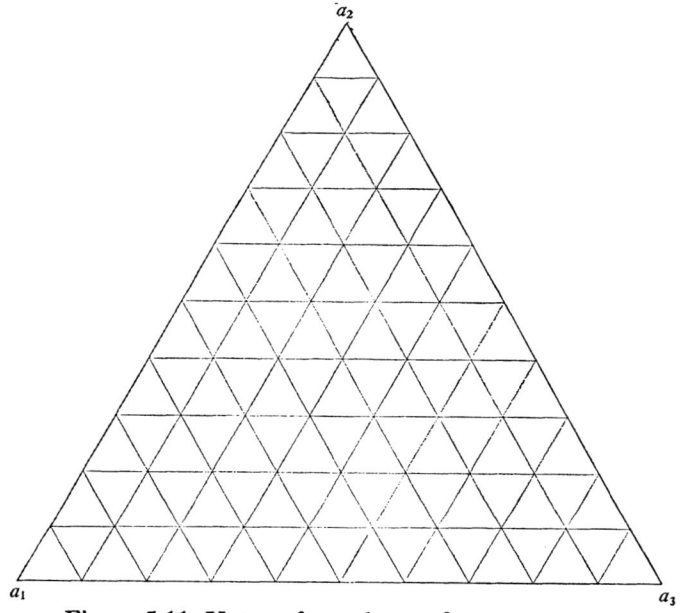

Figure 5.11. Untransformed gene-frequency space.

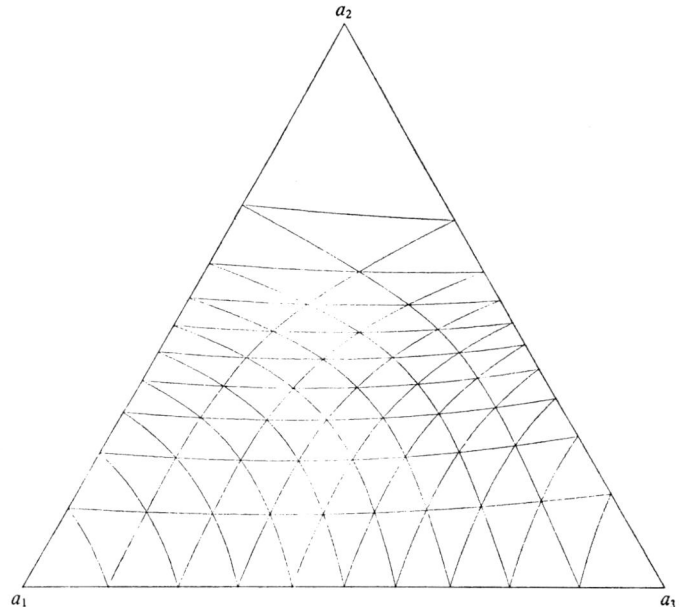

Figure 5.12. Transformed gene-frequency space (see table 5.1).

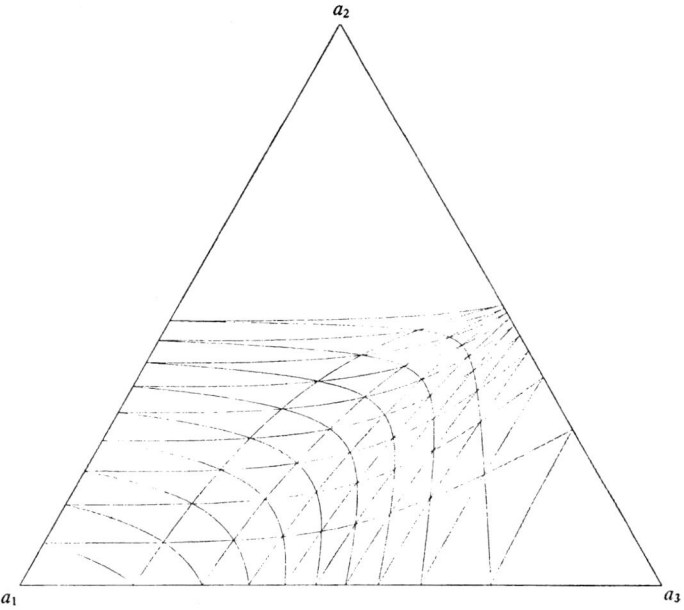

Figure 5.13. Transformed gene-frequency space (see table 5.1).

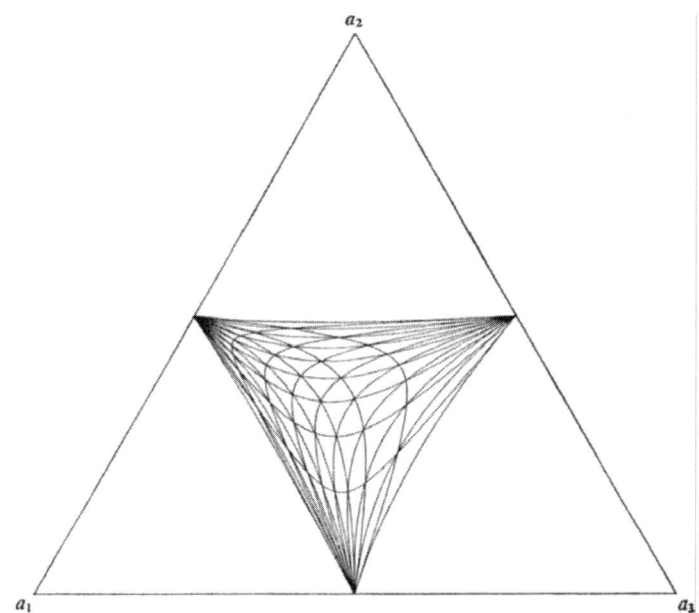

Figure 5.14. Transformed gene-frequency space (see table 5.1).

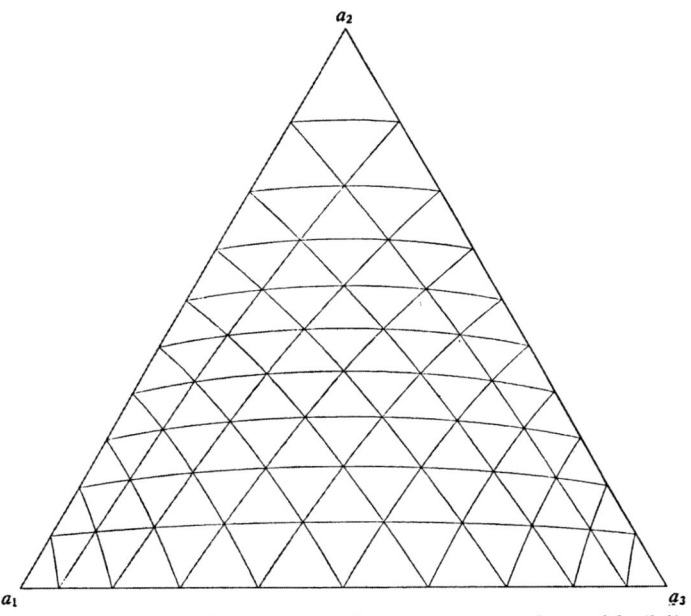

Figure 5.15. Transformed gene-frequency space (see table 5.1).

figure 5.6, at (0.1, 0.8, 0.1), and figure 5.10 to a population which started at the point (0.099, 0.8, 0.101) near the apex of figure 5.8.

Figures 5.12 to 5.15 show the mappings of the gene-frequency space (figure 5.11) as a result of one generation of selection, with the same viability matrices as for figures 5.5 to 5.8 respectively. Note that each is topologically similar to figure 5.11 except for figure 5.14, where the homozygotes are lethal so that a population on any side of the triangle must move, in one generation, to the mid-point of that side.

5.5   CARTESIAN EQUATION OF THE CONIC

The figures of section 5.4 were for the most part derived in a straightforward manner using the recurrence relations for the gene frequencies, (4.2.4), and the value of the mean viability, (5.2.1). But in order to plot the mean-viability contours directly (that is, without using a general contour-plotting program) it was necessary to find the dimensions and orientation of the conic for each value of w. Homogeneous coordinates are not well-adapted for such metrical considerations, nor are they suited to controlling a computer plotter. On both counts Cartesian coordinates are required.

For any given viability matrix the mean-viability contours are homothetic (theorem 5.2.1), so it is only necessary to find the centre, type (ellipse or hyperbola), eccentricity, and orientation of the conical contour for a particular w. The conics for other values of w follow immediately by virtue of the fact that w and the square of any linear dimension (such as the major axis of an ellipse) are linearly related, since along any line the graph of the mean viability is a parabola (theorem 5.2.4).

We use the fact that there is only one conic with a given centre which passes through three specified points; its parametric equation may thus be found from these data alone. For any figure, therefore, we choose a convenient w, find three convenient points on the corresponding conic, transform them into Cartesian coordinates, and thus find the parametric equation of the conic. Knowing the dimension of the major axis (for example) of this conic for the chosen w, and the fact that for $w = \hat{w}$ this length is zero, we can then derive the major axis for any w. We give the details only for the elliptical and hyperbolic cases; if the contours were parabolae or straight lines special treatment would be necessary.

To find three points, choose $w = w_{22}$ in (5.2.1) so that the conic

71

passes through the vertex $(0, 1, 0)$, thus establishing the first point. A convenient second point is where it cuts $p_1 = 0$ again, which is at a value of p_2 given by

$$\left.\begin{aligned} w_{22} &= w_{22} p_2^2 + w_{33} p_3^2 + 2w_{23} p_2 p_3 \\ p_3 &= 1 - p_2 \end{aligned}\right\} . \tag{5.5.1}$$

The solution other than $p_2 = 1$ is simply

$$p_2 = \frac{w_{33} - w_{22}}{w_{22} + w_{33} - 2w_{23}} = \frac{w_{22} - w_{33}}{\sigma_{23}} . \tag{5.5.2}$$

Similarly, the third point is where it cuts $p_3 = 0$ again, at

$$p_2 = \frac{w_{11} - w_{22}}{w_{22} + w_{11} - 2w_{12}} = \frac{w_{22} - w_{11}}{\sigma_{12}} . \tag{5.5.3}$$

If $w_{22} = w_{23}$ or $w_{22} = w_{12}$ we start with another vertex.

Transforming to Cartesian coordinates by

$$\left.\begin{aligned} x &= (p_2 + 2p_3)/\sqrt{3} \\ y &= p_2 \end{aligned}\right\} , \tag{5.5.4}$$

suppose the centre (see section 5.2) is (X, Y) and the three points (X_1, Y_1), (X_2, Y_2), (X_3, Y_3). Without loss of generality, put $(X, Y) = (0, 0)$; we can transform back to the true origin later.

The parametric equation of an ellipse centred at the origin is

$$\left.\begin{aligned} x \cos \alpha + y \sin \alpha &= a \cos \theta \\ -x \sin \alpha + y \cos \alpha &= b \sin \theta \end{aligned}\right\} , \tag{5.5.5}$$

where α is the angle between the x-axis and the axis of the ellipse in the first quadrant, a that semi-axis, b the other semi-axis, and θ the parameter $(0 \leqslant \theta < 2\pi)$.

There are three pairs of equations to solve for α, a and b, such as

$$\left.\begin{aligned} X_1 \cos \alpha + Y_1 \sin \alpha &= a \cos \theta_1 \\ -X_1 \sin \alpha + Y_1 \cos \alpha &= b \sin \theta_1 \end{aligned}\right\} . \tag{5.5.6}$$

Setting

$$\left.\begin{aligned} X_1 \cos \alpha + Y_1 \sin \alpha &= \lambda_1 \\ -X_1 \sin \alpha + Y_1 \cos \alpha &= \mu_1 \end{aligned}\right\} \tag{5.5.7}$$

etc., and eliminating θ in each of the first two pairs,

$$\left.\begin{array}{l} \dfrac{\lambda_1^2}{a^2} + \dfrac{\mu_1^2}{b^2} = 1 \\[2mm] \dfrac{\lambda_2^2}{a^2} + \dfrac{\mu_2^2}{b^2} = 1 \end{array}\right\}, \qquad (5.5.8)$$

and

whence

$$\left.\begin{array}{l} a^2 = \dfrac{\lambda_1^2 \mu_2^2 - \mu_1^2 \lambda_2^2}{\mu_2^2 - \mu_1^2} \\[3mm] b^2 = \dfrac{\mu_1^2 \lambda_2^2 - \lambda_1^2 \mu_2^2}{\lambda_2^2 - \lambda_1^2} \end{array}\right\}. \qquad (5.5.9)$$

Substitution of a^2 and b^2 in the third equation of type (5.5.8) leads to

$$\lambda_1^2(\mu_3^2 - \mu_2^2) + \lambda_2^2(\mu_1^2 - \mu_3^2) + \lambda_3^2(\mu_2^2 - \mu_1^2) = 0. \qquad (5.5.10)$$

Substituting for the λs and μs in terms of (5.5.7) leads, after some reduction, to

$$\tan 2\alpha = \dfrac{-[X_1^2(Y_2^2 - Y_3^2) + X_2^2(Y_3^2 - Y_1^2) + X_3^2(Y_1^2 - Y_2^2)]}{\begin{array}{l}[X_1 Y_1(X_2^2 + Y_2^2 - X_3^2 - Y_3^2) + X_2 Y_2(X_3^2 + Y_3^2 - X_1^2 - Y_1^2) \\ + X_3 Y_3(X_1^2 + Y_1^2 - X_2^2 - Y_2^2)]\end{array}}. \qquad (5.5.11)$$

Thus we know α, and, from (5.5.9), a and b. The parametric equation for the ellipse, referred again to the true origin is, from (5.5.5),

$$\left.\begin{array}{l} x = X + a\cos\theta\cos\alpha - b\sin\theta\sin\alpha \\ y = Y + a\cos\theta\sin\alpha + b\sin\theta\cos\alpha \end{array}\right\}. \qquad (5.5.12)$$

Since these values a and b are for an elliptical contour corresponding to mean viability w, for any other w, say w', the values a' and b' will be related to a and b by the pair of equations

$$\left.\begin{array}{l} \dfrac{a'^2}{\hat{w} - w'} = \dfrac{a^2}{\hat{w} - w} \\[3mm] \dfrac{b'^2}{\hat{w} - w'} = \dfrac{b^2}{\hat{w} - w} \end{array}\right\}. \qquad (5.5.13)$$

For the hyperbolic case the parametric equation comparable to (5.5.5) is

$$\left.\begin{array}{l} x\cos\alpha + y\sin\alpha = a\sec\theta \\ -x\sin\alpha + y\cos\alpha = b\tan\theta \end{array}\right\}. \qquad (5.5.14)$$

Since $\sec^2\theta - \tan^2\theta = 1$ the effect of this is to change the signs in (5.5.8), and hence to change the sign of b^2 in (5.5.9). Equivalently, we can proceed with the original equations and recognize a hyperbolic case when b^2 is negative: this is a signal to use (5.5.14) with b set equal to $\sqrt{-b^2}$. Similarly, if a^2 is negative, this is a signal to use

$$\left.\begin{array}{l} x\cos\alpha + y\sin\alpha = a\,\tan\theta \\ -x\sin\alpha + y\cos\alpha = b\,\sec\theta \end{array}\right\}, \qquad (5.5.15)$$

with a set equal to $\sqrt{-a^2}$, which is a member of the family of hyperbolae conjugate to (5.5.14). Since the equilibrium point is where the asymptotes intersect, (5.5.14) and (5.5.15) will cover the cases $w > \hat{w}$ and $w < \hat{w}$, though not necessarily respectively. In calculating the scaling factors (5.5.13), the denominators should always be taken positively.

The asymptotes are

$$x(a\sin\alpha \mp b\cos\alpha) = y(a\cos\alpha \pm b\sin\alpha)$$

referred to the centre, and the angle between them is $2\tan^{-1}(b/a)$.

5.6 THE LINES OF EQUAL AVERAGE EFFECT

In the preceding sections we have reproduced, in homogeneous form, most of the results given by Feller (1969), except his geometrical delimitation of the region within which the point representing the gene frequencies in the next generation must lie. In this section we merely note the properties of certain lines which are relevant to Feller's considerations, and which form the basis of further developments. Consider the equation for the centre of the conic:

$$w_{11}x + w_{12}y + w_{13}z$$
$$= w_{12}x + w_{22}y + w_{23}z = w_{13}x + w_{23}y + w_{33}z. \qquad (5.2.3\ bis)$$

We know that at the centre each part of the equation is equal to $\hat{w}$ (section 5.3), and hence that the three lines

$$\left.\begin{array}{l} w_{11}p_1 + w_{12}p_2 + w_{13}p_3 - \hat{w} = 0 \\ w_{12}p_1 + w_{22}p_2 + w_{23}p_3 - \hat{w} = 0 \\ w_{13}p_1 + w_{23}p_2 + w_{33}p_3 - \hat{w} = 0 \end{array}\right\} \qquad (5.6.1)$$

intersect at the centre (they may be rendered homogeneous by multiplying $\hat{w}$ by $(p_1 + p_2 + p_3)$). In terms of the average effects

(2.3.4) these equations may be written

$$\left.\begin{array}{c} \alpha_1 + w - \hat{w} = 0 \\ \alpha_2 + w - \hat{w} = 0 \\ \alpha_3 + w - \hat{w} = 0 \end{array}\right\}, \qquad (5.6.2)$$

where w is, as usual, the value of the mean viability at (p_1, p_2, p_3). Now

$$\frac{p_i'}{p_i} = \frac{\alpha_i + w}{w} \qquad (2.3.5 \ bis)$$

so that on the ith line

$$\frac{p_i'}{p_i} = \frac{\hat{w}}{w}. \qquad (5.6.3)$$

This shows that for any point on the ith line the frequency of the ith gene will change by a factor given by the ratio of the equilibrium mean viability to the present mean viability.

Furthermore, any linear combination of the equations (5.6.2) will be a line through the centre, and we may consider especially the three lines formed by pairwise subtraction:

$$\left.\begin{array}{c} \alpha_2 = \alpha_3 \\ \alpha_1 = \alpha_3 \\ \alpha_1 = \alpha_2 \end{array}\right\}. \qquad (5.6.4)$$

These lines have been considered by Kirkman (1967), and are the lines joining the centre to the 'modal points' of Feller (1969). They are the three lines along which the average effects, taken in pairs, are equal. The line $\alpha_2 = \alpha_3$, for example, passes through the centre and the equilibrium point (internal or external) on $p_1 = 0$; for when gene a_1 is absent the solution of $\alpha_2 = \alpha_3$ gives the equilibrium point for a_2 and a_3, as it must because the average effects for a_2 and a_3 will then be equal (and zero). Anywhere on this line, given (2.3.5 bis),

$$\frac{p_2'}{p_2} = \frac{p_3'}{p_3}$$

or

$$\frac{p_2'}{p_3'} = \frac{p_2}{p_3}. \qquad (5.6.5)$$

This shows that any point on the line $\alpha_2 = \alpha_3$ will lead to a point, in the next generation, on the line connecting it to the vertex $p_1 = 1$. The reader will find it instructive to cover figure 5.8 (especially) with tracing paper and sketch such lines.

CHAPTER 6

AN X-LINKED LOCUS

6.1 NO SELECTION

So far we have been treating an autosomal locus and it has not been necessary to consider the sexes of the individuals in the population, having made the assumptions that mating is between individuals of opposite sex and that any locus on an autosome segregates independently of sex. Sex is determined (in mammals) by a pair of chromosomes which are morphologically dissimilar, labelled X and Y. As with autosomes, every normal individual carries two members of this pair, but the only combinations which arise are XX and XY: for XX produces the female phenotype, and XY the male, so that every mating is of the type XX by XY, resulting in offspring of genotype XX and XY each with probability one half according to Mendel's First Law (section 1.1). Maternal gametes all carry an X-chromosome, whilst paternal gametes are of two equally frequent kinds, X-bearing and Y-bearing.

But segregation independent of sex will clearly not be the rule with any locus on the sex chromosomes X and Y. We will confine our attention to a locus on that part of the X-chromosome which has no counterpart on the Y-chromosome (the *differential region* of the X-chromosome; such a locus is said to exhibit *complete X-linkage*). The reason for this is that there is hardly any evidence in man for the presence of important loci either on the differential region of the Y-chromosome (leading to *complete Y-linkage*) or on the *homologous region* of the X- and Y-chromosomes (that is, the part which behaves, in segregation, like an autosome, but whose loci would exhibit *partial X-linkage* – which might equally be called partial Y-linkage – and would not segregate independently of sex).

Consider, then, a locus X with alleles x_1 and x_2 at initial frequencies p_f, q_f in the females and p_m, q_m in the males. The possible genotypes are $x_1 x_1$, $x_1 x_2$ and $x_2 x_2$ in the females and x_1 and x_2 in the males (since they can carry only one copy of a gene at the locus). If, in the notation of section 1.3, we label the female genotypes G_1, G_2 and G_3, respectively, and the male genotypes, x_1Y and x_2Y, G_4 and G_5, and regard Y as an allele at locus X, theorem 1.3.1 holds. The female genotypic proportions will be $(f_1, f_2, f_3, 0, 0)$ and the male $(0, 0, 0, m_4, m_5)$, each mating thus being between a male and a female. Henceforth we assume random union of gametes.

After one generation of random mating, however, an equilibrium population does not result unless the gene frequencies in the two sexes are the same, and thus there is no immediate equivalent to Hardy–Weinberg equilibrium (section 1.4). This is because the gene frequency amongst the males is equal to that amongst their mothers (who contribute all the X-chromosomes carried by the males), whilst the gene frequency amongst the females is the mean of the frequencies amongst their mothers and fathers (each of whom contributes half the X-chromosomes carried by the females). A damped oscillation results.

The offspring genotype frequencies are

$$\left.\begin{array}{l} \text{Females} \left\{\begin{array}{ll} x_1 x_1 & p_m p_t \\ x_1 x_2 & p_m q_t + q_m p_t \\ x_2 x_2 & q_m q_t \end{array}\right. \\[2em] \text{Males} \left\{\begin{array}{ll} x_1 & p_t \\ x_2 & q_t \end{array}\right. \end{array}\right\} \qquad (6.1.1)$$

and hence the new gene frequencies are

$$\left.\begin{array}{l} p_t' = \tfrac{1}{2}(p_m + p_t) \\ p_m' = p_t \end{array}\right\}. \qquad (6.1.2)$$

and

Writing $p = \tfrac{1}{3}(p_m + 2p_t)$, the overall frequency of gene x_1 on the assumption of equal numbers of males and females,

$$p' = \tfrac{1}{3}(p_m' + 2p_t') = \tfrac{1}{3}(p_m + 2p_t) = p, \qquad (6.1.3)$$

indicating that the overall gene frequency is unchanging from one generation to the next, as must indeed be the case because every X-chromosome, whether carried by a male or a female, has an equal chance of contributing to the next generation.

From (6.1.2),

$$p_m' - p_t' = -\tfrac{1}{2}(p_m - p_t), \qquad (6.1.4)$$

so that the difference in male and female gene frequencies is not immediately reduced to zero, as with an autosomal locus (section

2.1), but is halved in magnitude each generation:

$$p_{\mathrm{m}}^{[n]} - p_{\mathrm{f}}^{[n]} = (-\tfrac{1}{2})^n (p_{\mathrm{m}} - p_{\mathrm{f}}). \tag{6.1.5}$$

Thus $p_{\mathrm{m}}^{[n]}$ and $p_{\mathrm{f}}^{[n]}$ both converge on p; indeed

$$\left. \begin{aligned} p_{\mathrm{f}}^{[n]} - p &= (-\tfrac{1}{2})^n (p_{\mathrm{f}} - p) \\ p_{\mathrm{m}}^{[n]} - p &= (-\tfrac{1}{2})^n (p_{\mathrm{m}} - p) \end{aligned} \right\}. \tag{6.1.6}$$

With any number of alleles the equations (6.1.6) still hold for each allele separately, since from its point of view all the other alleles can be considered identical, when the case reduces to that of two alleles.

Note that (6.1.2) leads to

$$p_{\mathrm{f}}'' = \tfrac{1}{2}(p_{\mathrm{m}}' + p_{\mathrm{f}}') = \tfrac{1}{2}(p_{\mathrm{f}} + p_{\mathrm{f}}'), \tag{6.1.7}$$

so that the frequency of any allele in the females is the mean of its frequencies amongst the females of the preceding two generations. Since $p_{\mathrm{m}}' = p_{\mathrm{f}}$, the same is true for the males.

6.2 SELECTION WITH TWO ALLELES

Let the female genotypes $x_1 x_1$, $x_1 x_2$ and $x_2 x_2$ have viabilities w_{11}, w_{12} and w_{22} respectively, and the male genotypes x_1 and x_2 viabilities v_1 and v_2 respectively. Let the gametes that form the first generation of offspring have frequencies $p_{\mathrm{f}} x_1$ and $q_{\mathrm{f}} x_2$ from the females, and $p_{\mathrm{m}} x_1$ and $p_{\mathrm{m}} x_2$ from the males. Then the offspring genotypic frequencies, before selection, will be as given in (6.1.1), and after selection will be proportional to

$$\left. \begin{aligned} x_1 x_1 \quad & p_{\mathrm{m}} p_{\mathrm{f}} w_{11} \\ x_1 x_2 \quad & (p_{\mathrm{m}} q_{\mathrm{f}} + q_{\mathrm{m}} p_{\mathrm{f}}) w_{12} \\ x_2 x_2 \quad & q_{\mathrm{m}} q_{\mathrm{f}} w_{22} \end{aligned} \right\} \tag{6.2.1}$$

in the females and proportional to

$$\left. \begin{aligned} x_1 \quad & p_{\mathrm{f}} v_1 \\ x_2 \quad & q_{\mathrm{f}} v_2 \end{aligned} \right\} \tag{6.2.2}$$

in the males. We do not write out the constants of proportionality because, as we shall see, they are of little interest in selection at an X-linked locus, and it is preferable to work in the gene ratios

$u_f = p_f/q_f$ and $u_m = p_m/q_m$. We find:

Theorem 6.2.1. Recurrence relations for the gene ratios

The gene ratios in the next generation are given by

$$\left.\begin{aligned} u_f' &= \frac{2w_{11}u_fu_m + w_{12}(u_f+u_m)}{2w_{22}+w_{12}(u_f+u_m)} \\ u_m' &= \frac{v_1u_f}{v_2} \end{aligned}\right\} \quad ** \qquad (6.2.3)$$

We note that if u_f and u_m are positive, so are u_f' and u_m', and hence so are all subsequent values.

Thus in the next generation

$$u_f'' = \frac{2w_{11}v_1u_f'u_f + w_{12}(v_2u_f'+v_1u_f)}{2w_{22}v_2+w_{12}(v_2u_f'+v_1u_f)}. \qquad (6.2.4)$$

At gene-frequency equilibrium $u_f'' = u_f' = u_f$ (if this is possible) $= \hat{u}_f$, say, and

$$\hat{u}_f = \hat{u}_f \frac{2w_{11}v_1\hat{u}_f + w_{12}(v_1+v_2)}{2w_{22}v_2+w_{12}(v_1+v_2)\hat{u}_f}, \qquad (6.2.5)$$

whence

$$2w_{11}v_1\hat{u}_f + w_{12}(v_1+v_2) = 2w_{22}v_2 + w_{12}(v_1+v_2)\hat{u}_f;$$

or $$\hat{u}_f = 0;$$

or $$\hat{u}_f = \infty.$$

Thus we have

Theorem 6.2.2. Haldane's Theorem†

At a diallelic X-linked locus the internal equilibrium gene ratio in the females (if it exists) is

$$\hat{u}_f = \frac{\frac{1}{2}w_{12}(v_1+v_2)-w_{22}v_2}{\frac{1}{2}w_{12}(v_1+v_2)-w_{11}v_1}. \quad ** \qquad (6.2.6)$$

Since $0 \leqslant u \leqslant \infty$, there can only be equilibria in the permitted range $0 \leqslant p \leqslant 1$ if $\frac{1}{2}w_{12}(v_1+v_2)-w_{22}v_2$ and $\frac{1}{2}w_{12}(v_1+v_2)-w_{11}v_1$ are the same sign. Equation (6.2.6) is the same equation as that for an auto-

† See the Historical notes, section 6.4.

somal locus (2.2.4) when the viabilities of the genotypes $a_1 a_1$, $a_1 a_2$ and $a_2 a_2$ are $w_{11} v_1$, $\frac{1}{2} w_{12}(v_1 + v_2)$ and $w_{22} v_2$ respectively. Naturally, therefore, $\hat{p}_f \, (= \hat{u}_f/(1 + \hat{u}_f))$ is a stationary point of the function

$$w = w_{11} v_1 p_f^2 + w_{12}(v_1 + v_2) p_f q_f + w_{22} v_2 q_f^2, \qquad (6.2.7)$$

but it is wrong to conclude that this is an increasing function under selection; C. Cannings has provided the following counter-example: $w_{11} = 1$, $w_{12} = 2$, $w_{22} = 3$, $v_1 = 4$, $v_2 = 1$, $p_f = p_m = \frac{1}{2}$. With these initial gene frequencies $w = \frac{17}{4} = 4.25$, and in the next generation $p_f' = \frac{3}{8}$, $p_m' = \frac{4}{5}$ and $w' = \frac{261}{64} = 4.0781$. Cannings (1969 c) pointed out that if the equilibrium (6.2.6) is unstable there must nevertheless exist a set of gene frequencies which initiate convergence to it.

No increasing function is known for an X-linked locus, and it is for this reason that the constants of proportionality mentioned earlier, which are the female and male mean viabilities, are of little interest.

The equilibrium gene frequencies corresponding to (6.2.6) are

$$\left. \begin{aligned} \hat{p}_f &= \frac{\frac{1}{2} w_{12}(v_1 + v_2) - w_{22} v_2}{w_{12}(v_1 + v_2) - w_{11} v_1 - w_{22} v_2} \\[2mm] \hat{p}_m &= \frac{\frac{1}{2} w_{12} v_1(v_1 + v_2) - w_{22} v_1 v_2}{\frac{1}{2} w_{12}(v_1 + v_2)^2 - v_1 v_2(w_{11} + w_{22})} \end{aligned} \right\}. \qquad (6.2.8)$$

We now consider the stability of this equilibrium, and the question of convergence to the equilibrium when it is stable. We first prove

Theorem 6.2.3. Stability conditions for the internal equilibrium

The internal equilibrium given by Haldane's Theorem (theorem 6.2.2) is stable if and only if both numerator and denominator of (6.2.6) are positive.

Proof.† Without loss of generality, let $w_{12} = 2$ and $v_2 = 1 - v_1$, $0 \leqslant v_1, v_2 \leqslant 1$, so that the recurrence relations (6.2.3) become

$$\left. \begin{aligned} u_f' &= \frac{w_{11} u_f u_m + u_f + u_m}{w_{22} + u_f + u_m} \\[2mm] u_m' &= \frac{v_1 u_f}{1 - v_1} \end{aligned} \right\}. \qquad (6.2.9)$$

† See the Historical notes, section 6.4.

The necessary and sufficient condition for stability is that the roots of

$$\begin{vmatrix} \dfrac{\partial u'_\mathrm{f}}{\partial u_\mathrm{f}} - \lambda & \dfrac{\partial u'_\mathrm{f}}{\partial u_\mathrm{m}} \\[2mm] \dfrac{\partial u'_\mathrm{m}}{\partial u_\mathrm{f}} & \dfrac{\partial u'_\mathrm{m}}{\partial u_\mathrm{m}} - \lambda \end{vmatrix} = 0 \qquad (6.2.10)$$

should both lie in the interval ± 1, where the differential coefficients are the values at equilibrium

$$\left. \begin{aligned} \frac{\partial u'_\mathrm{f}}{\partial u_\mathrm{f}} &= \frac{w_{11}\hat{u}_\mathrm{m}^2 + w_{11}w_{22}\hat{u}_\mathrm{m} + w_{22}}{(w_{22} + \hat{u}_\mathrm{f} + \hat{u}_\mathrm{m})^2} \\[2mm] \frac{\partial u'_\mathrm{f}}{\partial u_\mathrm{m}} &= \frac{w_{11}\hat{u}_\mathrm{f}^2 + w_{11}w_{22}\hat{u}_\mathrm{f} + w_{22}}{(w_{22} + \hat{u}_\mathrm{f} + \hat{u}_\mathrm{m})^2} \\[2mm] \frac{\partial u'_\mathrm{m}}{\partial u_\mathrm{f}} &= \frac{v_1}{1 - v_1} \\[2mm] \frac{\partial u'_\mathrm{m}}{\partial u_\mathrm{m}} &= 0 \end{aligned} \right\} \qquad (6.2.11)$$

Thus (6.2.10) is

$$\lambda^2 - \frac{w_{11}\hat{u}_\mathrm{m}^2 + w_{11}w_{22}\hat{u}_\mathrm{m} + w_{22}}{(w_{22} + \hat{u}_\mathrm{f} + \hat{u}_\mathrm{m})^2}\lambda - \frac{v_1(w_{11}\hat{u}_\mathrm{f}^2 + w_{11}w_{22}\hat{u}_\mathrm{f} + w_{22})}{(1 - v_1)(w_{22} + \hat{u}_\mathrm{f} + \hat{u}_\mathrm{m})^2} = 0.$$
$$(6.2.12)$$

We now rewrite this as

$$\lambda^2 - L\lambda - M = 0,$$

where L and M are evidently positive, and use the easily proved result that both roots lie between ± 1 if and only if

$$L + M < 1.$$

With the simplification $w_{12} = 2$ and $v_2 = 1 - v_1$, (6.2.6) becomes

$$\left. \begin{aligned} \hat{u}_\mathrm{f} &= \frac{1 - w_{22}(1 - v_1)}{1 - w_{11}v_1} \\[2mm] \hat{u}_\mathrm{m} &= \frac{v_1}{1 - v_1}\hat{u}_\mathrm{f} \end{aligned} \right\} \qquad (6.2.13)$$

and we also have

81

Substituting $\hat{u}_m$ for $\frac{v_1}{1-v_1}\hat{u}_t$ in M,

$$L+M = \frac{w_{11}\hat{u}_m(w_{22}+\hat{u}_t+\hat{u}_m)+w_{22}\left(\frac{1}{1-v_1}+w_{11}\hat{u}_m\right)}{(w_{22}+\hat{u}_t+\hat{u}_m)^2}.$$

(6.2.14)

Now, taking the identity

$$(1-w_{11}v_1)\hat{u}_t = (1-v_1)\hat{u}_t+(1-w_{11})v_1\hat{u}_t$$

and using (6.2.13),

$$1-w_{22}(1-v_1) = (1-v_1)\hat{u}_t+(1-w_{11})(1-v_1)\hat{u}_m,$$

or

$$\frac{1}{1-v_1}+w_{11}\hat{u}_m = w_{22}+\hat{u}_t+\hat{u}_m.$$

Substituting this in (6.2.14),

$$L+M = \frac{w_{11}\hat{u}_m+w_{22}}{w_{22}+\hat{u}_t+\hat{u}_m}$$

and

$$L+M-1 = \frac{w_{11}\hat{u}_m-\hat{u}_t-\hat{u}_m}{w_{22}+\hat{u}_t+\hat{u}_m} = \frac{\hat{u}_m(w_{11}-1/v_1)}{w_{22}+\hat{u}_t+\hat{u}_m}. \quad (6.2.15)$$

We have two cases to consider: the numerator and denominator of (6.2.13) both positive, and both negative. In the first, $w_{11}v_1 < 1$ and thus $L+M-1 < 0$, $L+M < 1$, and the equilibrium is necessarily stable; in the second, $w_{11}v_1 > 1$ and $L+M-1 > 0$, $L+M > 1$, and the equilibrium is necessarily unstable. **

Using only the gene-ratio u_t, and discarding its subscript, we now consider the question of convergence when the equilibrium is stable, following Cannings (1967).

Theorem 6.2.4. Cannings' Theorem

If an internal equilibrium is stable, the gene ratios will converge to it from all internal starting points.

Proof. Consider the case

$$u = u' \geqslant \hat{u}. \quad (6.2.16)$$

Then since

$$u \geqslant \hat{u} = \frac{1-w_{22}(1-v_1)}{1-w_{11}v_1}$$

and both numerator and denominator are positive,

$$(1 - w_{11}v_1)u \geqslant [1 - w_{22}(1 - v_1)]$$

or $$u + w_{22}(1 - v_1) \geqslant w_{11}v_1 u + 1,$$

whence $$u \geqslant \frac{u(w_{11}v_1 u + 1)}{u + w_{22}(1 - v_1)}. \qquad (6.2.17)$$

Now consider (6.2.4) with $w_{12} = 2$, $v_2 = 1 - v_1$, and $u' = u$:

$$u'' = \frac{u(w_{11}v_1 u + 1)}{u + w_{22}(1 - v_1)},$$

so that (6.2.17) is $$u \geqslant u''. \qquad (6.2.18)$$

We next note that if (6.2.4) is written

$$u^{[n+2]} = f(u^{[n]}, u^{[n+1]})$$

then f is a real function of two positive real variables $u^{[n]}$ and $u^{[n+1]}$ on the positive half-line, and has continuous partial derivatives of all orders with respect to its variables. In particular, differentiation of (6.2.4) shows that both $\partial f/\partial u^{[n]}$ and $\partial f/\partial u^{[n+1]}$ are positive.

Thus, recalling (6.2.16) and using Taylor expansions,

$$\hat{u} = f(\hat{u}, \hat{u}) \leqslant f(u, u') = u''; \qquad (6.2.19)$$

and if $u^{[n]}, u^{[n+1]} \geqslant \hat{u}$,

$$u^{[n+2]} = f(u^{[n]}, u^{[n+1]}) \geqslant f(\hat{u}, \hat{u}) = \hat{u}; \qquad (6.2.20)$$

and if $u^{[n]} \geqslant u^{[n+1]} \geqslant u^{[n+2]}$,

$$u^{[n+3]} = f(u^{[n+1]}, u^{[n+2]}) \leqslant f(u^{[n]}, u^{[n+1]}) = u^{[n+2]}. \qquad (6.2.21)$$

Relations (6.2.18) to (6.2.21), used successively, imply that when $u = u' \geqslant \hat{u}$, (6.2.16), the sequence $\{u^{[n]}\}$ is a monotonic decreasing sequence with exact lower bound $\hat{u}$. Similar arguments lead to the result that when $u = u' \leqslant \hat{u}$, the sequence $\{u^{[n]}\}$ increases monotonically to $\hat{u}$.

Now consider a general sequence $\{u^{[n]}\}$ for which $u > u'$. Suppose $\{H^{[n]}\}$ and $\{K^{[n]}\}$ are two sequences of female gene ratios such that $H < K$ and $H' < K'$. Then by arguments similar to the above,

$H^{[n]} < K^{[n]}$ for all n. Now set $H = H' = u'$ and $K = K' = u$, then for all n, $K^{[n]} > u^{[n]} > H^{[n]}$. But sequences such as $\{H^{[n]}\}$ and $\{K^{[n]}\}$, with their first two members equal, have been shown to converge to $\hat{u}$, whence $\{u^{[n]}\}$ must converge to $\hat{u}$ since it is bounded above and below by $\{K^{[n]}\}$ and $\{H^{[n]}\}$ respectively. Similarly if $u < u'$. **

Cannings (1967) also discussed the behaviour of the sequences when the equilibrium is unstable. If $u, u' \geqslant \hat{u}$, the sequence converges to infinity; if $u, u' \leqslant \hat{u}$, the sequence converges to zero; but if u, u' are on either side of $\hat{u}$, the result is uncertain. However, as soon as two successive values of the gene ratio are on the same side of the equilibrium, the die is cast.

[*Note added in proof.* Palm (1974) has since proved that in all cases the population converges to an equilibrium state.]

6.3 SELECTION WITH MANY ALLELES

The behaviour of a multiallelic sex-linked locus with no selection has already been treated in section 6.1. Under selection, a matrix formulation is available similar to that used in chapter 4 for an autosomal multiallelic locus. To save the use of subscripts we use f and m for the female and male gene frequencies rather than p_f and p_m as hitherto.

Let F be the $k \times k$ diagonal matrix of the frequencies of the genes $x_1, x_2, ..., x_k$ in the females and M the similar matrix for the males. Let W be the symmetric matrix of female genotypic viabilities in which $w_{ij} (= w_{ji})$ gives the viability of the genotype $x_i x_j (\equiv x_j x_i)$, and let V be the diagonal matrix of male genotypic viabilities in which v_i gives the viability of x_i.

The array of female genotypic frequencies before selection is

$$FUM + MUF,$$

where U is the $k \times k$ matrix with every element unity, and of male genotypic frequencies is simply

$$IF,$$

where I is the identity matrix. After selection the arrays will be proportional to

$$\left. \begin{array}{c} FWM + MWF \\ VF \end{array} \right\} \tag{6.3.1}$$

and

respectively, analogously to the autosomal case. Thus we have

Theorem 6.3.1. *Recurrence relations for the gene frequencies*

The female and male gene frequencies in the next generation are given by

$$F'1 = \frac{(FWM + MWF)1}{1^{\mathrm{T}}(FWM + MWF)1}$$

and

$$M'1 = \frac{VF1}{1^{\mathrm{T}}VF1}$$

$$(6.3.2)$$

respectively. **

Thus

$$F''1 = \frac{(F'WM' + M'WF')1}{1^{\mathrm{T}}(F'WM' + M'WF')1}$$

$$= \frac{(F'WVF + FVWF')1}{1^{\mathrm{T}}(F'WVF + FVWF')1},$$

$$(6.3.3)$$

since the second equation of (6.3.2)

$$M'1 \propto VF1$$

can equally be written $M' \propto VF = FV$,

all the matrices being diagonal.

Theorem 6.3.2. *Internal equilibria*

The internal stationary values of the gene frequencies, if they exist, are given by

$$\hat{F}1 = \frac{(WV + VW)^{-1}1}{1^{\mathrm{T}}(WV + VW)^{-1}1}.$$

$$(6.3.4)$$

Proof. At gene-frequency equilibrium $F'' = F' = F = \hat{F}$ and (6.3.3) becomes

$$\hat{F}1 = \frac{\hat{F}(WV + VW)\hat{F}1}{1^{\mathrm{T}}\hat{F}(WV + VW)\hat{F}1},$$

$$(6.3.5)$$

which is simply the autosomal equation (4.2.4) with viability matrix $(WV + VW)$. Thus it has solution (6.3.4) which corresponds to (4.2.10). **

We note that theorem 6.3.2 only holds if $(WV + VW)$ is non-singular (see section 4.5). We will not pursue the question of the stability of the equilibrium except to record that a necessary condition for stability is that

$$\mathbf{1}^T \hat{F}(WV + VW)\hat{F}\mathbf{1}$$

should be a maximum at equilibrium (with respect to variation in $\hat{F}$).
Nothing is known about convergence.

6.4 HISTORICAL NOTES

The equations of section 6.1 are almost all due to Robbins (1918 a);
we have not attached his name to any theorem, however, since a
more important theorem in chapter 8 merits it. Haldane (1924) considered various special cases of the model for selection at a diallelic
locus (section 6.2). He introduced the general model in 1926, and
gave an approximate solution valid for weak selection. He derived
the equilibrium exactly, however, and gave, without proof, the
correct stability conditions.

Mandel (1959 a) pointed out the analogy with the autosomal case
which we have given. Kimura (1960) and Haldane and Jayakar
(1964) observed that Haldane's conditions were also those for
either gene to increase in frequency when introduced (at a low frequency) into a population consisting entirely of the other gene, but
this is not sufficient (as they thought) to prove the stability of the
intermediate equilibrium (Cannings, 1969 c).

I gave the first direct proof (Edwards, 1961), a simplified version
of which is given in section 6.2. Sprott (1957) considered the question of stability at a multiallelic locus, but did not arrive at explicit
conditions for two alleles; for many alleles the conditions he gave
require a knowledge of the equilibrium gene frequencies, which
was not available until the work of Cannings (1968 b). For section
6.3 we have used the approach of Cannings (1969 b).

Yaglom (1967) extended the concept of a de Finetti diagram
(chapter 3) to an X-linked diallelic locus, the analogous gene-
frequency space being a prism.

CHAPTER 7

MISCELLANEOUS SINGLE-LOCUS MODELS

7.1 VIABILITIES DIFFERENT IN THE TWO SEXES

In this chapter we consider briefly two models which arise when the simple single-autosomal-locus model is generalized; we start by relaxing the restriction that the viabilities of each genotype should be the same in each sex, and then, in section 7.2, consider a model in which the viabilities differ amongst several niches.

Although Haldane (1924) considered the case of selection in one sex only and (1926) the case of slow selection differing in males and females, the general diallelic model with six viability parameters, one for each genotype of each sex, was discussed without restriction by Owen (1952, 1953). Its chief interest lies in the fact that, for some choices of viabilities, it admits three equilibria, two of which may be stable. Thus the following scheme of viabilities

	Females	Males	
a_1a_1	0.5	2.15	
a_1a_2	1	1	(7.1.1)
a_2a_2	0.5	2.15	

generates three equilibria with gene ratios

	Females	Males	
1	0.5821	0.3389	
2	1	1	(7.1.2)
3	1.7179	2.9511	

of which Owen found the first and third to be stable, and the second unstable.

We shall not give a full algebraic treatment of Owen's model, because it has so far resisted attempts at a complete solution. Much of what is known about it has been found from a consideration of various special cases, of which Bodmer (1965) and Mandel (1971) give full accounts. Using procedures which are by now entirely standard, Owen, Bodmer and Mandel demonstrate that finding the equilibria necessitates solving a cubic equation, thus raising the possibility of there being three equilibria, a possibility realised by the above example. The stability of these equilibria is considered by using the standard method (see theorem 6.2.3) applied numerically; a general algebraic treatment is not available. Mandel does, however, obtain sufficient conditions for the existence of at least one internal equilibrium.

On the question of convergence, no increasing function of the gene frequencies is known, and C. Cannings has produced the following counter-example to the conjecture that the mean viability might be an increasing function, as in the autosomal case. Take as viabilities

	Females	Males	
a_1a_1	1	1	
a_1a_2	2	1	(7.1.3)
a_2a_2	1	1	

and as initial gametic frequencies for a_1, 0.75 from the females and 1 from the males. The mean viability (assuming equal numbers of females and males amongst the offspring) is $\frac{9}{8}$, and the gene frequencies in the next generation of gametes are $\frac{4}{5}$ and $\frac{7}{8}$ respectively. The mean viability of the corresponding population of offspring is found to be $\frac{91}{80}$. Mandel proved that near equilibrium the gene ratios change monotonically, thus eliminating the possibility of oscillations near the equilibria. Cannings (1969a) has proved that in the special case in which there is only selection in one sex (as in 7.1.3), the necessary and sufficient conditions for a stable equilibrium, and indeed the equilibrium itself, are exactly the same as in the model with the same viabilities in both sexes (Fisher's Theorem, theorem 2.2.1) and that these conditions are sufficient for convergence. For other special cases, the reader is referred to Bodmer (1965), Mandel (1971) and Li (1976).

In view of the lack of a complete algebraic treatment of a diallelic locus, it is not surprising that the general treatment of a multiallelic locus with sex-differential viabilities has proved intractable. Cannings (1969b) has given the matrix formulation using the parallel notation to that in chapter 4, and is able to make some progress when the male and female gene frequencies are equal at equilibrium, which he shows to be the case when there is selection in one sex only. The equilibrium is then the same as in the corresponding non-differential case (theorem 4.2.5) and he shows that sufficient conditions for the equilibrium to be stable are given by the conditions which are known to be both necessary and sufficient in the non-differential case (section 4.4).

7.2   VIABILITIES DIFFERENT IN SEVERAL NICHES

In this section we relax the assumption of a homogeneous environment, and suppose (following Levene, 1953) that, after random mating has produced a new generation of offspring, they migrate independently of genotype to m separate environments, or 'niches', where selection takes place, with viability w_{ijl} for genotype $a_i a_j$ in the lth niche. After selection, random mating occurs, a proportion c_l of the adults coming from the lth niche ($\Sigma c_l = 1$).

We find
$$p_i' = \sum_l (c_l \sum_j p_i p_j w_{ijl}/w_l), \quad \text{all } i, \tag{7.2.1}$$

where
$$w_l = \sum_{i,j} p_i p_j w_{ijl} \tag{7.2.2}$$

is the mean viability in the ith niche. At equilibrium (if such exist) $p_i' = p_i = \hat{p}_i$, say, all i, and from (7.2.1)

$$1 = \sum_l (c_l \sum_j \hat{p}_j w_{ijl}/\hat{w}_l), \quad \text{all } i, \tag{7.2.3}$$

where
$$\hat{w}_l = \sum_{i,j} \hat{p}_i \hat{p}_j w_{ijl}.$$

The solution of (7.2.3) is clearly difficult; with only two alleles and two niches it is a pair of equations which reduce first to a quartic in p_1, and then to a cubic since $(1-p_1)$ is necessarily a factor. Each additional niche increases the degree by two, and with more than two alleles simultaneous equations are involved as well.

We shall not give a general algebraic treatment, but confine ourselves to demonstrating that with three alleles a line equilibrium can occur, the example being due to Cannings (1971). Consider two niches of equal size ($c_1 = c_2 = \frac{1}{2}$), and viability matrices

$$
\begin{array}{c}
\text{niche 1} \quad \begin{pmatrix} 1 & 2 & 2 \\ 2 & 1 & 2 \\ 2 & 2 & 1 \end{pmatrix} \\[2em]
\text{niche 2} \quad \begin{pmatrix} 2 & 1 & 1 \\ 1 & 2 & 1 \\ 1 & 1 & 2 \end{pmatrix}
\end{array}
. \tag{7.2.4}
$$

Equation (7.2.3) is

$$2 = \frac{p_1 + 2p_2 + 2p_3}{p_1^2 + p_2^2 + p_3^2 + 4(p_2 p_3 + p_1 p_3 + p_1 p_2)}$$

$$+ \frac{2p_1 + p_2 + p_3}{2(p_1^2 + p_2^2 + p_3^2 + p_2 p_3 + p_1 p_3 + p_1 p_2)} \quad (7.2.5)$$

for $i = 1$, with similar equations (and identical denominators) for $i = 2$ and $i = 3$. Now $p_1 + p_2 + p_3 = 1$ and, taking its square,

$$2(p_2 p_3 + p_1 p_3 + p_1 p_2) = 1 - (p_1^2 + p_2^2 + p_3^2) = 1 - \Sigma, \quad \text{say,}$$

whence (7.2.5) is
$$2 = \frac{2 - p_1}{2 - \Sigma} + \frac{1 + p_1}{1 + \Sigma}, \quad (7.2.6)$$

and similarly for $i = 2$ and $i = 3$. It is immediately obvious that both $p_i = \Sigma$ and $\Sigma = \frac{1}{2}$ are solutions to all three equations, and it is a matter of simple algebra to show that there are no other solutions. $p_i = \Sigma$ (all i) implies $p_1 = p_2 = p_3 = \frac{1}{3}$, and $\Sigma = \frac{1}{2}$ implies the homogeneous form

$$p_1^2 + p_2^2 + p_3^2 - 2(p_2 p_3 + p_1 p_3 + p_1 p_2) = 0. \quad (7.2.7)$$

In homogeneous coordinates, the first is an equilibrium point at the centre of the triangle of reference (necessitated by symmetry, of course), whilst the second (7.2.7) is a line of equilibria, in fact the inscribed circle to the triangle.

In view of the general intractability of the recurrence relations and the equilibrium equations, it is remarkable that it is possible to find an increasing function of the gene frequencies, akin to the mean viability when the environment is homogeneous. Li (1955) considered the geometric mean of the niche mean viabilities

$$w^* = \prod_l w_l^{c_l}, \quad (7.2.8)$$

and Cannings (1971) pointed out that since

$$\ln w^* = \sum_l c_l \ln w_l,$$

$$\frac{1}{w^*} \frac{\partial w^*}{\partial p_i} = \sum_l \frac{c_l}{w_l} \frac{\partial w_l}{\partial p_i},$$

$$\frac{p_i}{2w^*} \frac{\partial w^*}{\partial p_i} = \sum_l \left(c_l \frac{p_i}{2} \frac{\partial w_l}{\partial p_i} \middle/ w_l \right) = p_i',$$

by (7.2.1), and therefore that, by Baum and Eagon's Theorem (theorem 4.7.1), we have

Theorem 7.2.1. An increasing function for the multi-niche model

In a multi-niche model with any number of alleles at a single locus, the geometric mean of the niche mean viabilities never decreases from one generation to the next, and remains constant if and only if the gene frequencies are at equilibrium values. **

We do not follow this with an Equivalence Theorem (compare theorem 4.4.1) because we have not excluded the possibility that a population may move from the vicinity of one local maximum to a higher point on a different local maximum.

Although a general algebraic treatment is not available, (7.2.8) provides us with a fairly clear picture of what to expect, for the increasing function is but a combination of the conic-sectioned hills, depressions and saddles provided by the mean-viability surface for each of the niches. A maximum point is therefore likely to be close to the maximum point for a particular niche, and similarly for a mini- mum point, and this will be especially true if the niche is a dominant one (c_l large). Indeed, it is difficult to visualize a maximum *point* which is not produced in this way. Saddle-points and line equilibria may be produced, however, by the interaction of the surfaces, as in Cannings' example (7.2.4), where a concentric hill (niche 1) and depression (niche 2) of identical sizes interact to form a radially symmetric 'volcano' whose section is shown in figure 7.1.

In the same notation as (7.2.6),

$$w^* = \sqrt{[(2-\Sigma)(1+\Sigma)]}. \tag{7.2.9}$$

Now the distance from any point (p_1, p_2, p_3) to the centre $(\frac{1}{3}, \frac{1}{3}, \frac{1}{3})$ is given by r where
$$r^2 = \tfrac{2}{3}(p_1^2 + p_2^2 + p_3^2 - \tfrac{1}{3})$$
$$= \tfrac{2}{3}(\Sigma - \tfrac{1}{3}).$$

Thus w^* may be written as a function of r:
$$w^* = \sqrt{(\tfrac{20}{9} + \tfrac{1}{2}r^2 - \tfrac{9}{4}r^4)},$$

which is the function plotted in figure 7.1. The surface is obviously radially symmetrical.

An alternative to Levene's model is one in which instead of a proportion c_l of the mature individuals coming from the *l*th niche

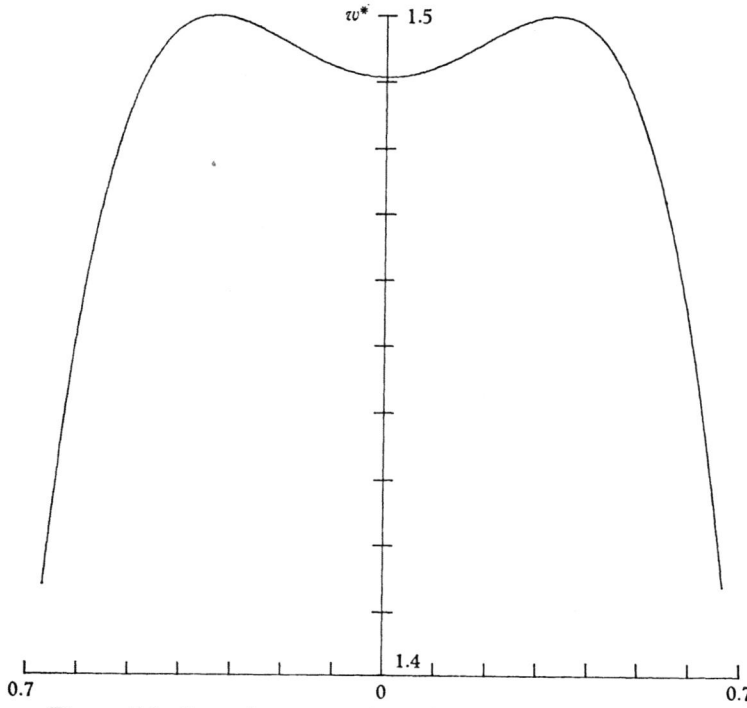

Figure 7.1. Central cross-section of the viability function w^*
in Cannings' example (7.2.4).

after selection, a proportion c_l of the offspring migrate to the lth
niche *before* selection (Dempster, 1955). The proportion coming
from that niche after selection will then be

$$b_l = c_l w_l / \sum_l c_l w_l, \qquad (7.2.10)$$

whence
$$p_i' = \sum_l (b_l \sum_j p_i p_j w_{ijl}/w_l), \quad \text{all } i, \qquad (7.2.11)$$

$$= \sum_l \sum_j p_i p_j c_l w_{ijl} / \sum_l c_l w_l$$

$$= \sum_j p_i p_j w_{ij}^* / \sum_{i,j} p_i p_j w_{ij}^*, \qquad (7.2.12)$$

where
$$w_{ij}^* = \sum_l c_l w_{ijl}. \qquad (7.2.13)$$

But (7.2.12) is simply the homogeneous model (cf. (2.3.2)) with
viabilities w_{ij}^* given by (7.2.13). It is therefore completely solved by
the methods of chapter 4.

If, in the models of this section, the adult individuals do not return to a central point to mate at random, but redistribute themselves amongst the niches to mate at random within each niche, the pattern of redistribution being governed by a migration matrix, the so-called 'migration models' result. We shall not consider these, and the reader is referred to Christiansen and Feldman (1975) and Christiansen (1975) for a survey and references. Similarly we shall not consider models for migrational selection, in which there is a characteristic migration matrix for each genotype (with or without genotypic selection as well). Suggested by Fisher (1930), these seem to have attracted little attention (Edwards, 1963; Parsons, 1963; Moody, 1981).

CHAPTER 8

TWO DIALLELIC LOCI

8.1 NO SELECTION

Two loci situated on different autosomes will, on the simplest genetic model, segregate independently, in which case Mendel's Second Law – of *independent segregation* – is said to hold. In the absence of selection, therefore, the two loci can be treated separately, and joint results obtained on the assumption of independence (under selection this will not be true in general, as will be shown in the next section).

Two loci situated on the same autosome may, however, exhibit the phenomenon of *linkage*, in which, by contrast to Mendel's Second Law, the genes carried on the same member of a pair of autosomes tend to stay together and enter the same gamete. In general, linkage is not complete, and its magnitude is measured by the *recombination fraction*, r, between the two loci.

Let two loci, A and B, possess alleles a_1, a_2 and b_1, b_2 respectively, and consider the genotype a_1b_1/a_2b_2, where the notation indicates that a_1 and b_1 are on one member of the pair of autosomes, and a_2 and b_2 on the other; a_1 and b_1 are then said to be in *coupling*, whereas a_1 and b_2 are in *repulsion*. The recombination fraction is defined to be the fraction of gametes in which recombination has taken place between the two loci during gametogenesis, giving rise, in the present example, to the gametes a_1b_2 and a_2b_1 as opposed to a_1b_1 and a_2b_2. Strictly, the 'recombinant' gametes will be produced whenever there is an odd number of recombinant events between the two loci. There is a rich theory which relates the recombination fraction to the *map distance* between the two loci, and a corresponding statistical theory which enables the recombination fraction, and hence the map distance, to be estimated from data on the segregation of genes at two loci, but we will not pursue it (see Bailey, 1961), nor will we describe the physical basis of recombination, which involves the breakage and rejoining of chromosome strands.

All that is necessary for a development of the theory of gene-frequency change is the extension of Mendel's First Law to incorporate the phenomenon of linkage. In the above example the frequencies of the four types of gametes will be

$$
\left.\begin{array}{cc}
a_1 b_1 & \tfrac{1}{2}(1-r) \\
a_1 b_2 & \tfrac{1}{2}r \\
a_2 b_1 & \tfrac{1}{2}r \\
a_2 b_2 & \tfrac{1}{2}(1-r)
\end{array}\right\}, \tag{8.1.1}
$$

in conformity with both the definition of the recombination fraction and Mendel's First Law for each locus. If one of the two loci is homozygous then there are, of course, only two types of gamete (each with frequency one half), and if both loci are homozygous, only one type. It is the existence of double heterozygotes that gives rise to the distinctive features of two-locus theory.

When $r = 0$ linkage is complete, and the four types of gamete can be treated as four alleles at a tetra-allelic locus, the methods of chapter 4 applying. When $r = \tfrac{1}{2}$ segregation is evidently independent, as though the loci were on separate autosomes. With the discovery of linkage, which was unknown to Mendel, his Second Law becomes tautological. It is possible to construct models for recombination in which r exceeds one half, and although this possibility is covered by the following treatment, r is generally thought of as lying between 0 and $\tfrac{1}{2}$, the two extremes corresponding to the special cases mentioned above.

Let the four types of gamete produced by one generation of parents have frequencies

$$
\left.\begin{array}{cc}
a_1 b_1 & c_1 \\
a_1 b_2 & c_2 \\
a_2 b_1 & c_3 \\
a_2 b_2 & c_4
\end{array}\right\}, \tag{8.1.2}
$$

where $c_1 + c_2 + c_3 + c_4 = 1$. We may assume the equivalence of random mating and random union of gametes because, from the point of view of theorem 1.3.1, the four gametes are simply four 'genes'. If we retain, for the moment, the distinction between maternal and paternal gametes, placing the former above the line and the latter below, random mating will lead to sixteen offspring types

$$
\frac{a_1 b_1}{a_1 b_1}, \ldots, \frac{a_2 b_2}{a_2 b_2}
$$

(which fall into just nine genotype classes, the three genotypes at one locus combining in all possible ways with the three at the other).

The frequencies of the sixteen offspring types will be given by the square of the array of gamete frequencies, or $(c_1+c_2+c_3+c_4)^2$. Only the double heterozygotes, as we have already noted, will produce gametes at different frequencies in the presence of linkage from those in the absence of linkage: a_1b_1/a_2b_2 and a_2b_2/a_1b_1, which together have frequency $2c_1c_4$, will each produce

$$\left.\begin{array}{ll} a_1b_1 & \frac{1}{2}(1-r) \\ a_1b_2 & \frac{1}{2}r \\ a_2b_1 & \frac{1}{2}r \\ a_2b_2 & \frac{1}{2}(1-r) \end{array}\right\}, \qquad (8.1.1\ bis)$$

whilst a_1b_2/a_2b_1 and a_2b_1/a_1b_2, which together have frequency $2c_2c_3$, will each produce

$$\left.\begin{array}{ll} a_1b_1 & \frac{1}{2}r \\ a_1b_2 & \frac{1}{2}(1-r) \\ a_2b_1 & \frac{1}{2}(1-r) \\ a_2b_2 & \frac{1}{2}r \end{array}\right\}. \qquad (8.1.3)$$

Thus the gametic frequencies in the next generation (c_1', c_2', c_3', c_4') will be the same as those in the previous generation except for the change $(-c_1c_4r,\ c_1c_4r,\ c_1c_4r,\ -c_1c_4r)$ deriving from (8.1.1 bis) and the change $(c_2c_3r,\ -c_2c_3r,\ -c_2c_3r,\ c_2c_3r)$ deriving from (8.1.3). In all,

$$\left.\begin{array}{l} c_1' = c_1-r(c_1c_4-c_2c_3) \\ c_2' = c_2+r(c_1c_4-c_2c_3) \\ c_3' = c_3+r(c_1c_4-c_2c_3) \\ c_4' = c_4-r(c_1c_4-c_2c_3) \end{array}\right\}. \qquad (8.1.4)$$

Theorem 8.1.1. Robbins' Theorem, part I

At a pair of linked diallelic loci, the function of the gamete frequencies $(c_1c_4-c_2c_3)$ decreases by a factor $(1-r)$ each generation, where r is the recombination fraction between the two loci.

Proof. It follows from (8.1.4) that

$$(c_1'c_4'-c_2'c_3') = (1-r)(c_1c_4-c_2c_3). \qquad **$$

The coefficient $\Delta = c_1 c_4 - c_2 c_3$, or, in general,

$$\Delta^{[n]} = c_1^{[n]} c_4^{[n]} - c_2^{[n]} c_3^{[n]},$$

has acquired a large number of different names in the literature, not all of them appropriate; we shall call it *Robbins' coefficient*, after Robbins (1918*b*), to whom this theorem and the next are due. In the limit, after many generations, the alleles at each locus will be independently distributed in the population ($\Delta^{[\infty]} = 0$, provided $r > 0$), and we may say that the population is then in *Robbins equilibrium* (cf. *Hardy–Weinberg equilibrium*, section 1.4). It is interesting to note that Robbins' coefficient can be expressed as a linear combination of functions of the gamete frequencies: for

$$c_1 - c_2 - c_3 + c_4 = (c_1 - c_2 - c_3 + c_4)(c_1 + c_2 + c_3 + c_4)$$
$$= c_1^2 - c_2^2 - c_3^2 + c_4^2 + 2c_1 c_4 - 2c_2 c_3,$$

whence

$$2\Delta = c_1(1 - c_1) - c_2(1 - c_2) - c_3(1 - c_3) + c_4(1 - c_4).$$

The relationships between the gametic frequencies and the corresponding gene frequencies, $(c_1 + c_2)$, etc., are also worth noting:

$$\left.\begin{aligned}
c_1 &= (c_1 + c_2)(c_1 + c_3) + \Delta\\
c_2 &= (c_1 + c_2)(c_2 + c_4) - \Delta\\
c_3 &= (c_3 + c_4)(c_1 + c_3) - \Delta\\
c_4 &= (c_3 + c_4)(c_2 + c_4) + \Delta
\end{aligned}\right\} . \qquad (8.1.5)$$

The first equation of (8.1.4) can also be written

$$c_1' = (1 - r)c_1 + r(c_1 + c_2)(c_1 + c_3),$$

or $\quad c_1' - (c_1 + c_2)(c_1 + c_3) = [1 - r][c_1 - (c_1 + c_2)(c_1 + c_3)],$

and similarly for the others. Now from (8.1.4) it is clear that the individual gene frequencies, $(c_1 + c_2)$, $(c_1 + c_3)$, $(c_2 + c_4)$, $(c_3 + c_4)$, are invariants $((c_1' + c_2') = (c_1 + c_2)$, etc.) as must be the case with no selection. Thus

$$c_1^{[n]} - (c_1 + c_2)(c_1 + c_3) = [1 - r][c_1^{[n-1]} - (c_1 + c_2)(c_1 + c_3)],$$

whence clearly

$$c_1^{[n]} - (c_1 + c_2)(c_1 + c_3) = [1 - r]^n [c_1 - (c_1 + c_2)(c_1 + c_3)], \qquad (8.1.6)$$

and similarly for the other gamete frequencies. This form, due to

Malécot (1948, 1969), shows that the departure of each gamete frequency from its stationary value is reduced by a factor $(1-r)$ each generation (for $r > 0$), the stationary value being simply the gamete frequency under independent segregation $(c_1^{[\infty]} = (c_1+c_2)(c_1+c_3)$, from (8.1.6)). Alternatively, and equivalently, we can easily express (8.1.6) in terms of the departure of each gamete frequency from its initial frequency; this was the form in which Robbins originally expressed the results:

Theorem 8.1.2. Robbins' Theorem, part II

At a pair of linked diallelic loci, the gametic frequencies in the nth generation are given by

$$\left.\begin{aligned}
c_1^{[n]} &= c_1 - [1-(1-r)^n]\Delta \\
c_2^{[n]} &= c_2 + [1-(1-r)^n]\Delta \\
c_3^{[n]} &= c_3 + [1-(1-r)^n]\Delta \\
c_4^{[n]} &= c_4 - [1-(1-r)^n]\Delta
\end{aligned}\right\} . \quad ** \qquad (8.1.7)$$

The equilibrium result $c_1^{[\infty]} = c_1 - \Delta$ is, of course, the same as that already given above in a different form.

Since from part II of Robbins' Theorem the genotypic frequencies resulting from the random union of gametes can be obtained, it supplies the extension to the Hardy–Weinberg Theorem necessary for two linked loci. It is readily extended to the case of two loci with any number of alleles at each (indeed, Malécot's treatment is general in this respect), but the extension to an arbitrary number of loci is somewhat complex: pairwise independence at equilibrium is assured by the above results, but there remain all the higher-order interactions to consider. That these, too, are reduced each generation has been proved for three loci by Geiringer (1945) and for an arbitrary number by Bennett (1954).

8.2 SELECTION

We shall not attempt a comprehensive account of the theory of selection at two linked loci, such as Arunachalam and Owen (1971) have attempted, nor will we succumb to the temptation to proceed by way of numerical examples: rather, we hope to give some insight into the two-locus case by establishing certain theorems for partially restricted models. It is an unfortunate fact that the case of independent segregation ($r = \frac{1}{2}$) is hardly any simpler than the general case with recombination fraction r, with which we therefore start. Let the viabilities for each of the nine genotypes be as follows:

	$b_1 b_1$	$b_1 b_2$	$b_2 b_2$	
$a_1 a_1$	w_{11}	w_{12}	w_{13}	
$a_1 a_2$	w_{21}	w_{22}	w_{23}	(8.2.1)
$a_2 a_2$	w_{31}	w_{32}	w_{33}	

Note that, in order to avoid the proliferation of subscripts, w_{ij} no longer has its former meaning, and w_{ij} is not necessarily equal to w_{ji}. The mean viability is

$$w = w_{11}c_1^2 + w_{13}c_2^2 + w_{31}c_3^2 + w_{33}c_4^2$$
$$+ 2(w_{12}c_1 c_2 + w_{21}c_1 c_3 + w_{22}c_1 c_4$$
$$+ w_{22}c_2 c_3 + w_{23}c_2 c_4 + w_{32}c_3 c_4) \quad (8.2.2)$$

and the recurrence relations easily seen to be

$$\left.\begin{aligned}
wc_1' &= w_{11}c_1^2 + w_{12}c_1 c_2 + w_{21}c_1 c_3 + w_{22}c_1 c_4 - rw_{22}\Delta \\
wc_2' &= w_{12}c_1 c_2 + w_{13}c_2^2 + w_{22}c_2 c_3 + w_{23}c_2 c_4 + rw_{22}\Delta \\
wc_3' &= w_{21}c_1 c_3 + w_{22}c_2 c_3 + w_{31}c_3^2 + w_{32}c_3 c_4 + rw_{22}\Delta \\
wc_4' &= w_{22}c_1 c_4 + w_{23}c_2 c_4 + w_{32}c_3 c_4 + w_{33}c_4^2 - rw_{22}\Delta
\end{aligned}\right\} . \quad (8.2.3)$$

These equations may be looked upon either as three independent equations describing the change in gamete frequencies (c_1, c_2, c_3) from one generation to the next (remembering $c_4 = 1 - c_1 - c_2 - c_3$), or as three independent equations describing the changes in the gene frequencies ($c_1 + c_2$) and ($c_1 + c_3$) and in the Robbins coefficient Δ. From the latter point of view, there will be an equilibrium if the gene frequencies and Δ are at stationary values.

Fisher (1930) must have had these equations clearly in mind when he wrote the following passage:

Two factors, the alternative genes in which may be represented by A, a and B, b will maintain each other mutually in genetic equilibrium, if the selective advantage of A over a is reversible when B is substituted for b, or vice versa. Without attempting to specify the exact selective advantage enjoyed by each of the nine genotypes we may specify the type of selection under consideration by saying that A is advantageous in the presence of B but disadvantageous in the presence of b, and that B is advantageous in the presence of A but disadvantageous in the presence of *a*. Equally of course in this statement we might transpose the words advantageous and disadvantageous.

Equilibrium in such a system evidently implies that the increase in the frequency of *A* which takes place in the presence of *B* shall be exactly counterbalanced by its decrease in the presence of *b*; and that the increase in *B* which takes place in the presence of *A* shall be exactly counterbalanced by its decrease in the presence of *a*. But it is important to notice that the equilibrium of the frequencies of the gametic combinations *AB*, *Ab*, *aB*, *ab* requires a third condition of equilibrium. By the conditions of our problem, two of these, which we have chosen to be *AB* and *ab*, are favoured by Natural Selection, and increase in their zygotic stages, while the opposite pair *Ab* and *aB* decrease. The adjustment of the ratio between the frequencies of these two pairs of gametic types must take place by recombination in those individuals which are heterozygotes for both factors. Of these so-called double heterozygotes some arise by the union of the gametic types *AB* and *ab*, and in these the effect of recombination is to diminish the frequencies of these two types. This effect will be partially counteracted by recombination in heterozygotes of the second kind, arising from the union of *Ab* and *aB*; and, if the net effect of recombination is to decrease the frequencies of *AB* and *ab*, it is obvious that double heterozygotes derived from gametes of these kinds must be the more numerous.

Haldane (1931, 1932) must also have had such equations in mind when he wrote down the recurrence relations for the gene frequencies for the case a_1 fully dominant to a_2 and b_1 fully dominant to b_2, but he assumed $\Delta = 0$, no doubt on the basis of Robbins' Theorem, part I (theorem 8.1.1), which he had reproduced in Haldane (1926) in apparent ignorance of Robbins' work. Haldane assumed small selective coefficients and used the differential-equation approximation, concluding that Fisher's scheme was not sufficient to ensure stable equilibrium. Perhaps it was this paper which induced Fisher to replace 'will' by 'may' in the first sentence quoted above when revising his book (Fisher, 1958). Haldane (1931) also considered the case of each heterozygote's viability being the arithmetic mean of the corresponding homozygotes'.

Equations (8.2.3) were first given explicitly by Lewontin and Kojima (1960), although Kimura (1956) gave their continuous-time analogues. For more detailed historical comments, see the review by Bodmer and Parsons (1962).

Very little is known about the general system defined by these equations; not even the equilibria. The mean viability w (8.2.2) is not, in general, an increasing function of the gene frequencies, even with unlinked loci, as was first shown numerically by Kojima and Kelleher (1961), who also showed when, on the basis of Kimura's continuous-time model, such a result might be expected. Moran (1964) investigated the matter in some detail and proved the theorem which we name after him (theorem 8.2.1), from which this behaviour of the mean viability is evident since a stationary point of the function w in the space of gamete frequencies will not necessarily correspond to an equilibrium of the system (8.2.3). For w is the same function as is obtained in a tetra-allelic case with gene frequencies (c_1, c_2, c_3, c_4) and viability matrix

$$\begin{pmatrix} w_{11} & w_{12} & w_{21} & w_{22} \\ w_{12} & w_{13} & w_{22} & w_{23} \\ w_{21} & w_{22} & w_{31} & w_{32} \\ w_{22} & w_{23} & w_{32} & w_{33} \end{pmatrix}, \tag{8.2.4}$$

and therefore has its internal stationary point(s) (if any) at $(c_1^*, c_2^*, c_3^*, c_4^*)$ where

$$\left.\begin{aligned} w_{11}c_1^* + w_{12}c_2^* + w_{21}c_3^* + w_{22}c_4^* &= w^* \\ w_{12}c_1^* + w_{13}c_2^* + w_{22}c_3^* + w_{23}c_4^* &= w^* \\ w_{21}c_1^* + w_{22}c_2^* + w_{31}c_3^* + w_{32}c_4^* &= w^* \\ w_{22}c_1^* + w_{23}c_2^* + w_{32}c_3^* + w_{33}c_4^* &= w^* \end{aligned}\right\}, \tag{8.2.5}$$

by (4.2.9), writing w^* rather than $\hat{w}$ for the equilibrium mean viability of this tetra-allelic system. This enables us to state and prove

Theorem 8.2.1. Moran's Theorem

An internal equilibrium point for two linked diallelic loci is a stationary point for the mean viability in the space of gamete frequencies if and only if $rw_{22}\Delta$ is zero.

Proof. Suppose an internal equilibrium point is given by $\hat{c}_i = c_i^*$, $i = 1, 2, 3, 4$ (and thus $\hat{\Delta} = \Delta^*$). Then, inserting these values in (8.2.3) at its equilibrium $c_i' = c_i = \hat{c}_i$, $i = 1, 2, 3, 4$,

$$\left.\begin{aligned} \hat{w}c_1^* &= w^*c_1^* - rw_{22}\Delta^* \\ \hat{w}c_2^* &= w^*c_2^* + rw_{22}\Delta^* \\ \hat{w}c_3^* &= w^*c_3^* + rw_{22}\Delta^* \\ \hat{w}c_4^* &= w^*c_4^* - rw_{22}\Delta^* \end{aligned}\right\}, \tag{8.2.6}$$

and by addition $\hat{w} = w^*$. Therefore $rw_{22}\Delta^* = 0$ is a necessary condition.

Now suppose we have an internal equilibrium point such that $rw_{22}\hat{\Delta} = 0$; the equilibrium equations, from (8.2.3), are of the form

$$\hat{w}\hat{c}_1 = \hat{c}_1(w_{11}\hat{c}_1 + w_{12}\hat{c}_2 + w_{21}\hat{c}_3 + w_{22}\hat{c}_4),$$

with three similar equations for the other gamete frequencies. Since the point is internal, we recover the equations (8.2.5). Hence $rw_{22}\hat{\Delta} = 0$ is a sufficient condition. **

We now show that the mean viability is an increasing function when either (a) $r = 0$; (b) $w_{22} = 0$; (c) the viability matrix is additive over loci; or (d) the viability matrix is multiplicative over loci *and* initially $\Delta = 0$. We restrict our treatment to these cases.

(a) is simply the tetra-allelic case, covered by theorem 4.3.2, and so, in essence, is (b). For when $w_{22} = 0$ the value of r has no effect on the gamete frequencies in the next generation (8.2.3), nor on the mean viability (8.2.2). So we may as well put $r = 0$, which is (a). The equilibrium is given by the solution to (8.2.5).

The proof of (c) uses the same idea, and is due to Ewens (1969):

Theorem 8.2.2. *Ewens' Theorem*

For two diallelic loci with viabilities additive over loci, the mean viability increases from generation to generation except at equilibrium, when it remains the same.

Proof. Let the viabilities be of the form

$$w_{ij} = u_i + v_j. \tag{8.2.7}$$

In this case u_1, u_2 and u_3 are the viabilities of the three genotypes at the A locus, and v_1, v_2 and v_3 the similar viabilities at the B locus, and they combine additively. Then, from (8.2.2),

$$w = u_1(c_1 + c_2)^2 + 2u_2(c_1 + c_2)(c_3 + c_4) + u_3(c_3 + c_4)^2$$
$$+ v_1(c_1 + c_3)^2 + 2v_2(c_1 + c_3)(c_2 + c_4) + v_3(c_2 + c_4)^2. \tag{8.2.8}$$

Now $(c_1 + c_2)$, etc., are the frequencies of the individual genes, and w depends only on them, and not on the recombination fraction r. In the next generation, w' is similarly only dependent on the new gene frequencies $(c_1' + c_2')$, etc., which are themselves independent of r, by (8.2.3). Thus $\Delta w = w' - w$ is independent of r, and is consequently unchanged on putting $r = 0$; but then there is no recombination, and the four gametes act as four alleles, so that $\Delta w > 0$ except at equilibrium, when $\Delta w = 0$, by theorem 4.3.2. **

To complete the treatment of this additive case it only remains to find the equilibrium. We know that the equilibrium mean viability $\hat{w}$ will be that of the corresponding tetra-allelic system, by the above considerations; in that system the equilibrium is given by (8.2.5), but now we have $w_{ij} = u_i + v_j$, and writing $\hat{w}$ for w^*,

$$\alpha_1 = c_1^* + c_2^*, \quad \alpha_2 = c_3^* + c_4^*,$$

$$\beta_1 = c_1^* + c_3^* \quad \text{and} \quad \beta_2 = c_2^* + c_4^*,$$

we find

$$
\begin{aligned}
(1) \quad & u_1\alpha_1 + u_2\alpha_2 + v_1\beta_1 + v_2\beta_2 = \hat{w} \\
(2) \quad & u_1\alpha_1 + u_2\alpha_2 + v_2\beta_1 + v_3\beta_2 = \hat{w} \\
(3) \quad & u_2\alpha_1 + u_3\alpha_2 + v_1\beta_1 + v_2\beta_2 = \hat{w} \\
(4) \quad & u_2\alpha_1 + u_3\alpha_2 + v_2\beta_1 + v_3\beta_2 = \hat{w}
\end{aligned}
\right\} . \quad (8.2.9)
$$

These equations evidently do not have a unique solution, the matrix of viabilities being singular; indeed $(1) + (4) = (2) + (3)$. But $(1) - (3)$ gives
$$\alpha_1(u_1 - u_2) + \alpha_2(u_2 - u_3) = 0,$$
and $(1) - (2)$ gives $\beta_1(v_1 - v_2) + \beta_2(v_2 - v_3) = 0,$
leading to the well-known equilibrium gene ratios

$$
\left.
\begin{aligned}
\alpha_1/\alpha_2 &= (u_2 - u_3)/(u_2 - u_1) \\
\beta_1/\beta_2 &= (v_2 - v_3)/(v_2 - v_1)
\end{aligned}
\right\} . \quad (8.2.10)
$$

As we observed in section 2.2, both numerator and denominator must be of the same sign for each equilibrium to be internal.

Now (8.2.10) is not a point equilibrium, but a line in the gene-frequency space. Indeed, as may easily be shown, it is the line $w = \hat{w}$. We know by Ewens' Theorem that our two-locus population reaches this line in the space of gamete frequencies, the mean viability function (8.2.8) being exactly the same in both spaces. Thus (8.2.10) gives the gene frequencies at equilibrium in the two-locus problem. To find the gamete frequencies at equilibrium we note that, from (8.2.3), the equation for $\hat{c}_1$ is

$$\hat{w}\hat{c}_1 = \hat{c}_1(u_1\alpha_1 + u_2\alpha_2 + v_1\beta_1 + v_2\beta_2) - rw_{22}\hat{\Delta}$$

which, by (8.2.9), is $\hat{w}\hat{c}_1 = \hat{w}\hat{c}_1 - rw_{22}\hat{\Delta}.$

Since we are not assuming $r = 0$ or $w_{22} = 0$, $\hat{\Delta} = 0$. Thus we have

Theorem 8.2.3. Equilibrium for the additive viability model

For two loci with additive viabilities, an internal equilibrium exists if internal equilibria exist for each of the loci considered separately; the equilibrium gene frequencies will be as for the separate loci; and the Robbins coefficient will be zero. **

Karlin and Feldman (1970 a) have investigated the stability of the above equilibrium, and show that there are no surprises: when $r > 0$, the conditions $u_2 > u_1$, u_3 and $v_2 > v_1$, v_3 (which are, by Fisher's Theorem, theorem 2.2.1, the conditions for a stable internal equilibrium at each locus taken separately) ensure global convergence from any internal point. Bodmer and Felsenstein (1967) had earlier proved that these conditions ensure stability.

Case (d) is that of multiplicative viabilities of the form $w_{ij} = u_i v_j$. We suppose that initially $\Delta = 0$. The recurrence relations (8.2.3) are then

$$\left.\begin{aligned}
wc_1' &= c_1[u_1(v_1 c_1 + v_2 c_2) + u_2(v_1 c_3 + v_2 c_4)] \\
wc_2' &= c_2[u_1(v_2 c_1 + v_3 c_2) + u_2(v_2 c_3 + v_3 c_4)] \\
wc_3' &= c_3[u_2(v_1 c_1 + v_2 c_2) + u_3(v_1 c_3 + v_2 c_4)] \\
wc_4' &= c_4[u_2(v_2 c_1 + v_3 c_2) + u_3(v_2 c_3 + v_3 c_4)]
\end{aligned}\right\}, \qquad (8.2.11)$$

where $c_1 c_4 = c_2 c_3$.

We now prove the following theorem:

Theorem 8.2.4. Linked loci with multiplicative viabilities

Two diallelic loci with viabilities multiplicative over loci and gamete frequencies in Robbins equilibrium may be considered as independent loci.

Proof. Suppose the theorem true. Then, by (2.2.3),

$$\frac{c_1'}{c_2'} = \frac{c_1}{c_2} \cdot \frac{v_1 c_1 + v_2 c_2}{v_2 c_1 + v_3 c_2} = \frac{c_3'}{c_4'} = \frac{c_3}{c_4} \cdot \frac{v_1 c_3 + v_2 c_4}{v_2 c_3 + v_3 c_4} \qquad (8.2.12)$$

and, since

$$\frac{c_1}{c_2} = \frac{c_3}{c_4},$$

$$\frac{v_1 c_1 + v_2 c_2}{v_2 c_1 + v_3 c_2} = \frac{v_1 c_3 + v_2 c_4}{v_2 c_3 + v_3 c_4}. \qquad (8.2.13)$$

From (8.2.13)

$$\frac{u_1 + u_2 \dfrac{v_1 c_3 + v_2 c_4}{v_1 c_1 + v_2 c_2}}{u_1 + u_2 \dfrac{v_2 c_3 + v_3 c_4}{v_2 c_1 + v_3 c_2}} = 1,$$

whence the left equation of (8.2.12) may be written

$$\frac{c_1'}{c_2'} = \frac{c_1}{c_2} \cdot \frac{u_1(v_1 c_1 + v_2 c_2) + u_2(v_1 c_3 + v_2 c_4)}{u_1(v_2 c_1 + v_3 c_2) + u_2(v_2 c_3 + v_3 c_4)}. \tag{8.2.14}$$

But this is precisely the value of c_1'/c_2' obtained from (8.2.11). The other gamete ratios give similar results, and the theorem is proved. **

Since the mean viability of a population in Robbins equilibrium will be the product of the locus mean viabilities, it must be an increasing function. Alternatively, since Robbins equilibrium is maintained, the recurrence relations (8.2.11) never involve r, so we might as well put $r = 0$ throughout. The model is then identical to the corresponding tetra-allelic model, and again the mean viability must be an increasing function. Indeed, there is an isomorphism between the two models, but since theorem 8.2.4 completely solves the multiplicative-viability model in Robbins equilibrium, we will pursue the matter no further.

We should, however, note that nothing has been said about the possibility of equilibria with non-zero Robbins coefficient, and it is fairly obvious that if linkage is tight a small departure from $\Delta = 0$ will be magnified in the subsequent generation. Bodmer and Felsenstein (1967) show that a sufficient condition for the equilibrium $\Delta = 0$ to be stable is

$$r > \frac{(u_2 - u_1)(u_2 - u_3)}{u_2(2u_2 - u_1 - u_3)} \cdot \frac{(v_2 - v_1)(v_2 - v_3)}{v_2(2v_2 - v_1 - v_3)},$$

whilst Moran (1968) has shown that a sufficient condition for convergence is

$$\tfrac{1}{2} - r < \frac{(u_2 - u_1)(u_2 - u_3)}{6u_2(2u_2 - u_1 - u_3)}, \quad \frac{(v_2 - v_1)(v_2 - v_3)}{6v_2(2v_2 - v_1 - v_3)}.$$

This completes our account of two linked diallelic loci. The literature contains many treatments of models restricted in ways other than those we have described, perhaps the most detailed being the analysis by Karlin and Feldman (1970b) of the symmetric-viability model $w_{ij} = w_{ji}$, for which they show that there can be as many as seven equilibria. For further discussion of this and other special cases, Arunachalam and Owen (1971), Kimura and Ohta (1971), and Karlin (1975) may be consulted.

To make any substantial progress on the model with general viabilities, approximations are necessary. In this way the so-called 'equilibrium models' arise which have been widely discussed in the literature.

CHAPTER 9

FISHER'S FUNDAMENTAL THEOREM

9.1 THE FUNDAMENTAL THEOREM OF NATURAL SELECTION

In chapter 4 we learnt that the change in mean viability at a multiallelic locus from one generation to the next is

$$\Delta w = \frac{v}{w} + \mathbf{1}^{T}(\Delta P)W(\Delta P)\mathbf{1}$$

(Theorem 4.3.1, Li's Theorem). If there is no dominance in an additive sense then the matrix of viabilities W may be written $x\mathbf{1}^{T} + \mathbf{1}x^{T}$ (in the notation of section 4.5) whence

$$\Delta w = \frac{v}{w} + \mathbf{1}^{T}(\Delta P)(x\mathbf{1}^{T} + \mathbf{1}x^{T})(\Delta P)\mathbf{1}$$

$$= \frac{v}{w} + \mathbf{1}^{T}(\Delta P)(\mathbf{1}x^{T} + x\mathbf{1}^{T})(\Delta P)\mathbf{1}$$

$$= \frac{v}{w}$$

since $\mathbf{1}^{T}(\Delta P)\mathbf{1} = 0$. (A similar argument with multiplicative viabilities $W = xx^{T}$ shows that $\mathbf{1}^{T}(\Delta P)W(\Delta P)\mathbf{1}$ is then a squared quantity, leading to $\Delta w \geqslant v/w$.)

When the viabilities are additive we are in effect dealing with a population of genes whose association in genotypes is of no consequence to the theory, and the result $\Delta w = v/w$ has an interpretation far beyond the confines of mathematical genetics. It states that if a population of any kind is subdivided into groups each with its own growth-rate, defined as the factor by which it grows in a given time interval, then the change in the growth-rate of the population in this interval is equal to the variance in growth-rates amongst the groups, divided by the population growth-rate. The groups might be groups of animals or plants, but they might just as easily be economic sectors or stocks and shares. (A growth factor may of course be less than one, indicating shrinkage.) This simple theorem, which the reader should prove directly as an exercise, is the basis of Fisher's Fundamental Theorem of Natural Selection published in 1930. It may be called *Fisher's growth-rate theorem*, since it is not recorded any earlier (Edwards, 1994; see this paper for the simple proof).

The Fundamental Theorem adapts the growth-rate theorem to the genetical situation in which viabilities are not, in general, additive. Fisher (1930) first presented it as a continuous-time model, but Ewens (1989) developed a discrete-generation version which enabled the subtleties of the theorem to be clearly understood for the first time. In particular this showed that it is indeed not necessary to assume random mating and Hardy–Weinberg equilibrium, provided that the mating scheme is such as to ensure that the gene frequencies in the daughter generation at the time of conception are those in the adult population of parents. Since this more general treatment closely follows Fisher's development we no longer assume random mating as in earlier chapters. The following account is taken from Edwards (1994).

Theorem 9.1. The Fundamental Theorem of Natural Selection

The change in the mean viability ascribable solely to natural selection acting through changes in gene frequencies is exactly equal to the genetic variance[†] in viability, divided by the mean viability.

Proof. Let P_{ii} be the frequency of genotype a_ia_i, and $2P_{ij}$ the frequency of genotype a_ia_j. The frequency of a_i is then $p_i = \Sigma_j P_{ij}$ and the mean viability is $w = \Sigma_{i,j} P_{ij}w_{ij}$. In the next generation the frequency of a_i will therefore be $p' = \Sigma_j P_{ij}w_{ij}/w$.

We now need to find the average excess of gene a_i as defined in section 2.3. A minor problem of notation here intrudes itself, for the standard symbol for the average excess of the ith gene is a_i, which we have hitherto used for the gene itself. However, in this account we can avoid naming the genes explicitly again, allowing us to use a_i henceforth for the average excess. In accordance with the definition, in respect of viability

$$a_i = \frac{P_{ii}w_{ii} + \Sigma_{i,j,i \neq j}P_{ij}w_{ij}}{\Sigma_j P_{ij}} - w = \frac{\Sigma_j P_{ij}w_{ij}}{p_i} - w. \tag{9.1.1}$$

[†] Genic variance in modern terminology, see the preface.

The change in the frequency of the ith gene may thus be written

$$\Delta p_i = \frac{\Sigma_j P_{ij} w_{ij}}{w} - p_i = \left(\frac{\Sigma_j P_{ij} w_{ij}}{p_i} - w\right)\left(\frac{p_i}{w}\right) = \frac{a_i p_i}{w}, \qquad (9.1.2)$$

the same as it was under Hardy–Weinberg equilibrium (equation 2.3.10, recalling that α_i there stands both for the average excess and the average effect, equal under the Hardy–Weinberg assumption).

Next we need to find the average effect α_i of the ith gene as defined in section 2.3. In the absence of random mating this is no longer equal to the average excess. By definition the α_i are the values of α_i^* (say) which minimise the residual sum of squares $\Sigma_{i,j} P_{ij}(w - w_{ij} + \alpha_i^* + \alpha_j^*)^2$. The normal equations are, for all i,

$$\Sigma_j P_{ij}(w - w_{ij} + \alpha_i + \alpha_j) = 0,$$

and on expanding this and recalling that $\Sigma_j P_{ij} w_{ij} = p_i a_i + p_i w$ from (9.1.1), we find

$$p_i \alpha_i + \Sigma_j P_{ij} \alpha_j = p_i a_i. \qquad (9.1.3)$$

The genetic variance v is the sum of squares removed by the regression:

$$v = \Sigma_{i,j} P_{ij}(\alpha_i + \alpha_j)^2 = 2\Sigma_{i,j} P_{ij}\alpha_i^2 + 2\Sigma_{i,j} P_{ij}\alpha_i\alpha_j$$

$$= 2\Sigma_i p_i \alpha_i^2 + 2\Sigma_{i,j} P_{ij}\alpha_i\alpha_j.$$

Now $\qquad\qquad \Sigma_j P_{ij}\alpha_i\alpha_j = p_i(a_i - \alpha_i)\alpha_i$ from (9.1.3)

so $\qquad\qquad 2\Sigma_{i,j} P_{ij}\alpha_i\alpha_j = 2\Sigma_i p_i(a_i - \alpha_i)\alpha_i$

and we arrive at $\qquad\qquad v = 2\Sigma_i p_i a_i \alpha_i, \qquad (9.1.4)$

Fisher's expression for the genetic variance in terms of the average effects α_i and average excesses a_i of the genes (Fisher, 1930, though the formulation involving multiple alleles is 1958).

The final part of the proof of the theorem is to show that the *partial* change in the mean viability w due to changes in the genotypic frequencies P_{ij} from one generation to the next (P'_{ij}) is precisely equal to v/w. It is a standard result of the regression model that the weighted mean departure of the genotypic viabilities from their regression values is zero, or $0 = \Sigma_{i,j} P_{ij}(\alpha_i + \alpha_j)$, and adding the mean viability w to both sides of this leads to

$w = \Sigma_{i,j}P_{ij}(w + \alpha_i + \alpha_j)$ since $\Sigma_{i,j}P_{ij} = 1$. Thus the change in w ascribable solely to the change in the genotype frequencies P_{ij}, and not including changes in w and the α_i, is

$$\delta w = \Sigma_{i,j}(P'_{ij} - P_{ij})(w + \alpha_i + \alpha_j)$$
$$= \Sigma_{i,j}(P'_{ij} - P_{ij})(\alpha_i + \alpha_j)$$
$$= 2\Sigma_i(\alpha_i\Sigma_j(P'_{ij} - P_{ij}))$$
$$= 2\Sigma_i\alpha_i\Delta p_i \text{ since } \Sigma_jP_{ij} = p_i$$
$$= \frac{2\Sigma_i p_i a_i \alpha_i}{w} \text{ by } (9.1.2)$$
$$= \frac{v}{w} \text{ by } (9.1.4).$$

Thus the partial change in the mean viability w due solely to changes in the genotypic frequencies P_{ij} from one generation to the next is exactly equal to v/w. However, the third last equation above shows that the genotype frequencies are not explicitly involved, but only the gene-frequency changes. **

It is essential to understand the difference between Li's Theorem (4.3.1) and the Fundamental Theorem, between the total change in the mean viability Δw and the partial change δw, a difference which would exist in general even if Hardy–Weinberg equilibrium obtained. It is wholly attributable to the fact that in the next generation the mean fitness and the average effects will have changed slightly, changes with which the Fundamental Theorem is deliberately not concerned. The reason for this will be clearest to those familiar with the concept of *breeding value* in plant and animal breeding (see Falconer, 1989). The evolutionary arguments for this formulation cannot be aired in a book primarily concerned with an accurate rendering of the mathematics, but the reader who has worked through the above proof should be in a good position to follow Fisher's discourse in *The Genetical Theory of Natural Selection* (1930).

9.2 HISTORICAL NOTES

The Fundamental Theorem of Natural Selection has been widely misunderstood and the name frequently misappropriated for naive interpretations of it which are not even true. It is an important quantitative statement about the relation between variability and the efficacy of natural selection whose influence can be seen by the discerning investigator throughout evolutionary biology and artificial selection theory. It is 'fundamental' in the same sense that the Fundamental Theorem of Arithmetic is fundamental ('an integer can be expressed as a product of primes in one way only'), and quite possibly Fisher had this unconsciously in mind when naming his theorem.

For an extended account of the background to the theorem and of its subsequent history, and a comprehensive survey of the related literature, see Edwards (1994). For further developments and interpretations see Ewens (1992), Narain (1993), Frank (1997) and Lessard (1997).

REFERENCES

1866

Mendel, G. *Experiments in Plant Hybridisation* (Mendel's original paper in English translation, with a commentary by R. A. Fisher). Edinburgh: Oliver and Boyd, 1965.

1908

Hardy, G. H. Mendelian proportions in a mixed population. *Science*, **28**, 49–50.

Weinberg, W. Über den Nachweis der Vererbung beim Menschen. *Jahreshefte Verein f. vaterl. Naturk. in Württemberg*, **64**, 368–82.

1918

Fisher, R. A. The correlation between relatives on the supposition of Mendelian inheritance. *Trans. Roy. Soc. Edin.* **52**, 399–433. Reprinted, with a commentary by P. A. P. Moran and C. A. B. Smith, as *Eugenics Laboratory Memoirs* **41**, Cambridge University Press, 1966.

a Robbins, R. B. Applications of mathematics to breeding problems. II. *Genetics*, **3**, 73–92.

b Robbins, R. B. Some applications of mathematics to breeding problems. III. *Genetics*, **3**, 375–89.

1922

Fisher, R. A. On the dominance ratio. *Proc. Roy. Soc. Edin.* **42**, 321–41.

1924

Haldane, J. B. S. A mathematical theory of natural and artificial selection. Part I. *Trans. Camb. Phil. Soc.* **23**, 19–41.

1926

de Finetti, B. Considerazioni matematiche sull'ereditarietà mendeliana. *Metron*, **6**, 3–41.

Haldane, J. B. S. A mathematical theory of natural and artificial selection. Part III. *Proc. Camb. Phil. Soc.* **23**, 363–72.

Streng, O. Eine Völkerkarte. *Acta Soc. Med. Fennicae Duodecim*, **8**.

1930

Fisher, R. A. *The Genetical Theory of Natural Selection*. Oxford: Clarendon Press. Second edition: New York: Dover, 1958.

REFERENCES

1931
Haldane, J. B. S. A mathematical theory of natural selection. Part VIII. *Proc. Camb. Phil. Soc.* **27**, 137–42.

1932
Haldane, J. B. S. *The Causes of Evolution.* London: Longmans, Green; Ithaca, New York: Cornell University, 1966.

1937
Wright, S. The distribution of gene frequencies in populations. *Proc. Nat. Acad. Sci., Wash.* **23**, 307–20.

1941
Fisher, R. A. Average excess and average effect of a gene substitution. *Ann. Eugen.* **11**, 53–63.

1945
Geiringer, H. Further remarks on linkage in Mendelian heredity. *Ann. Math. Statist.* **16**, 390–3.

1948
Malécot, G. *Les mathématiques de l'hérédité.* Paris: Masson. (For an English translation, see Malécot, 1969.)

1952
Owen, A. R. G. A genetical system admitting of two stable equilibria. *Nature, Lond.* **170**, 1127.

1953
Levene, H. Genetic equilibrium when more than one ecological niche is available. *Amer. Nat.* **87**, 331–3.
Owen, A. R. G. A genetical system admitting of two distinct stable equilibria under natural selection. *Heredity.* **7**, 97–102.

1954
Bennett, J. H. On the theory of random mating. *Ann. Eugen.* **18**, 311–17.
Owen, A. R. G. Balanced polymorphism of a multiple allelic series. *Caryologia*, suppl. to **6**, 1240–1.

1955
Dempster, E. R. Maintenance of genetic heterogeneity. *Cold Spring Harbor Symp. Quant. Biol.* **20**, 25–32.
Li, C. C. The stability of an equilibrium and the average fitness of a population. *Amer. Nat.* **89**, 281–95.

1956
Kimura, M. A model of a genetic system which leads to closer linkage by natural selection. *Evolution*, **10**, 278–87.

1957
Kempthorne, O. *An Introduction to Genetic Statistics.* New York: Wiley.
Sprott, D. A. The stability of a sex-linked allelic system. *Ann. Hum. Genet., Lond.* **22**, 1–6.

REFERENCES

1958
Fisher, R. A. See Fisher (1930).
Kimura, M. On the change of population fitness by natural selection. *Heredity*, **12**, 145–67.
Mandel, S. P. H. and Hughes, I. M. Change in mean viability at a multiallelic locus in a population under random mating. *Nature, Lond.* **182**, 63–4.

1959
a Mandel, S. P. H. Stable equilibrium at a sex-linked locus. *Nature, Lond.* **183**, 1347–8.
b Mandel, S. P. H. The stability of a multiple allelic system. *Heredity*, **13**, 289–302.
Mulholland, H. P. and Smith, C. A. B. An inequality arising in genetical theory. *Amer. Math. Monthly*, **66**, 673–83.
Owen, A. R. G. Mathematical models for selection. In *Natural Selection in Human Populations*, ed. D. F. Roberts and G. A. Harrison. London: Pergamon.
Scheuer, P. A. G. and Mandel, S. P. H. An inequality in population genetics. *Heredity*, **13**, 519–24.

1960
Atkinson, F. V., Watterson, G. A. and Moran, P. A. P. A matrix inequality. *Quart. J. Math.* (2), **11**, 137–40.
Kimura, M. *Outline of Population Genetics*. Tokyo: Baifukan (in Japanese).
Lewontin, R. C. and Kojima, K. The evolutionary dynamics of complex polymorphisms. *Evolution*, **14**, 458–72.

1961
Bailey, N. T. J. *Introduction to the Mathematical Theory of Genetic Linkage*. Oxford: Clarendon Press.
Edwards, A. W. F. The population genetics of 'sex ratio' in *Drosophila pseudoobscura*. *Heredity*, **16**, 291–304.
a Kingman, J. F. C. A mathematical problem in population genetics. *Proc. Camb. Phil. Soc.* **57**, 574–82.
b Kingman, J. F. C. On an inequality in partial averages. *Quart. J. Math.* (2), **12**, 78–80.
Kojima, K. and Kelleher, T. M. Changes of mean fitness in random mating populations when epistasis and linkage are present. *Genetics*, **46**, 527–40.

1962
Bodmer, W. F. and Parsons, P. A. Linkage and recombination in evolution. *Adv. Genet.* **11**, 1–100.

1963
Edwards, A. W. F. Migrational selection. *Heredity.* **18**, 101–6.
Parsons, P. A. Migration as a factor in natural selection. *Genetica*, **33**, 184–206.

REFERENCES

1964

Falconer, D. S. *Introduction to Quantitative Genetics.* Edinburgh: Oliver and Boyd.

Haldane, J. B. S. and Jayakar, S. D. Equilibria under natural selection at a sex-linked locus. *J. Genet.* **59**, 29–36.

Moran, P. A. P. On the nonexistence of adaptive topographies. *Ann. Hum. Genet., Lond.* **27**, 383–93.

1965

Bodmer, W. F. Differential fertility in population genetics models. *Genetics*, **51**, 411–24.

1966

Tallis, G. M. Equilibria under selection for *k* alleles. *Biometrics*, **22**, 121–7.

1967

Baum, L. E. and Eagon, J. A. An inequality with applications to statistical estimation for probabilistic functions of Markov processes and to a model for ecology. *Bull. Amer. Math. Soc.* **73**, 360–3.

Blakley, G. R. Darwinian natural selection acting within populations. *J. theor. Biol.* **17**, 252–81.

Bodmer, W. F. and Felsenstein, J. Linkage and selection: theoretical analysis of the deterministic two locus random mating model. *Genetics*, **57**, 237–65.

Cannings, C. Equilibrium, convergence and stability at a sex-linked locus under natural selection. *Genetics*, **56**, 613–18.

Edwards, A. W. F. Review of *Commentary on R. A. Fisher's paper on the Correlation between Relatives on the Supposition of Mendelian Inheritance*, by P. A. P. Moran and C. A. B. Smith. *Ann. Hum. Genet., Lond.* **30**, 404–5.

Kingman, J. F. C. An inequality involving Radon–Nikodym derivatives. *Proc. Camb. Phil. Soc.* **63**, 195–8.

Kirkman, H. N. Haemoglobin abnormalities and stability of tri-allelic systems. *Ann. Hum. Genet., Lond.* **31**, 167–71.

Li, C. C. Fundamental theorem of natural selection. *Nature, Lond.* **214**, 505–6.

Yaglom, I. M. Geometric models of certain genetic processes. *Canad. J. Math.* **19**, 1233–42.

1968

a Cannings, C. Discrete generation models in population genetics. Ph.D. dissertation, University of London.

b Cannings, C. Equilibrium under selection at a multi-allelic sex-linked locus. *Biometrics*, **24**, 187–9.

Cannings, C. and Edwards, A. W. F. Natural selection and the deFinetti diagram. *Ann. Hum. Genet., Lond.* **31**, 421–8.

Moran, P. A. P. On the theory of selection dependent on two loci. *Ann. Hum. Genet., Lond.* **32**, 183–90.

REFERENCES
1969
a Cannings, C. Unisexual selection at an autosomal locus. *Genetics*, **62**, 225–9.
b Cannings, C. The study of multiallelic genetic systems by matrix methods. *Genet. Res., Camb.* **14**, 167–83.
c Cannings, C. A note on stability and convergence of genetic systems. *Evolution*, **23**, 517–18.
Ewens, W. J. A generalized fundamental theorem of natural selection. *Genetics*, **63**, 531–7.
Feller, W. A geometrical analysis of fitness in triply allelic systems. *Math. Biosciences*, **5**, 19–38.
Li, C. C. Increment of average fitness for multiple alleles. *Proc. Nat. Acad. Sci., Wash.* **62**, 395–8.
Malécot, G. *The Mathematics of Heredity.* San Francisco: Freeman.
Renwick, J. H. Progress in mapping human autosomes. *Br. med. Bull.* **25**, 65–73.
Wright, S. *Evolution and the Genetics of Populations*, volume 2: *The Theory of Gene Frequencies.* University of Chicago Press.

1970
Crow, J. F. and Kimura, M. *An Introduction to Population Genetics Theory.* New York: Harper and Row.
a Karlin, S. and Feldman, M. W. Convergence to equilibrium of the two locus additive viability model. *J. Appl. Prob.* **7**, 262–71.
b Karlin, S. and Feldman, M. W. Linkage and selection: two locus symmetric viability model. *Theor. Population Biol.* **1**, 39–71.
Kempthorne, O. and Pollack, E. Concepts of fitness in Mendelian populations. *Genetics*, **64**, 125–45.
Mandel, S. P. H. The equivalence of different sets of stability conditions for multiple allelic systems. *Biometrics*, **26**, 840–5.

1971
Arunachalam, V. and Owen, A. R. G. *Polymorphisms with Linked Loci.* London: Chapman and Hall.
Cannings, C. Natural selection at a multiallelic autosomal locus with multiple niches. *J. Genet.* **60**, 255–9.
Edwards, A. W. F. Review of *Evolution and the Genetics of Populations*, volume 2: *The Theory of Gene Frequencies*, by S. Wright. *Heredity*, **26**, 332–8.
Elandt-Johnson, R. C. *Probability Models and Statistical Methods in Genetics.* New York: Wiley.
Kimura, M. and Ohta, T. *Theoretical Aspects of Population Genetics.* Princeton University Press.
Mandel, S. P. H. Owen's model of a genetical system with differential viability between sexes. *Heredity*, **26**, 49–63.

1973

Seneta, E. On a genetic inequality. *Biometrics*, **29**, 810–14.

1974

Edwards, A. W. F. On Kimura's maximum principle in the genetical theory of natural selection. *Adv. Appl. Prob.* **6**, 1–3.

Palm, G. On the selection model for a sex-linked locus. *J. Math. Biol.* **1**, 47–50.

1975

Christiansen, F. B. Hard and soft selection in a sub-divided population. *Amer. Nat.* **109**, 11–16.

Christiansen, F. B. and Feldman, M. W. Subdivided populations: a review of the one- and two-locus deterministic theory. *Theor. Population Biol.* **7**, 13–38.

Hughes, P. J. and Seneta, E. Selection equilibria in a multiallele single-locus setting. *Heredity*, **35**, 185–94.

Ineichen, R. and Batschelet, E. Genetic selection and de Finetti diagrams. *J. Math. Biol.* **2**, 33–9.

Karlin, S. General two-locus selection models: some objectives, results and interpretations. *Theor. Population Biol.* **7**, 364–98.

1976

Li, C. C. *First Course in Population Genetics*. Pacific Grove, California: Boxwood.

1977

Edwards, A.W.F. Selection at a multiallelic locus: Feller's transformation. *Ann. Hum. Genet., Lond.* **41**, 219–24.

1981

Moody, M. Polymorphism with selection and genotype-dependent migration. *J. Math. Biol.* **11**, 245–67.

1989

Ewens, W.J. An interpretation and proof of the fundamental theorem of natural selection. *Theor. Population Biol.* **36**, 167–80.

Falconer, D.S. *Introduction to Quantitative Genetics*. Third edition. Harlow: Longman.

1992

Ewens, W.J. An optimizing principle of natural selection in evolutionary population genetics. *Theor. Population Biol.* **42**, 333–46.

1993

Narain, P. On an extremum principle in the genetical theory of natural selection. *J. Genet.* **72**, 59–71.

REFERENCES

1994

Edwards, A.W.F. The fundamental theorem of natural selection. *Biol. Rev.* **69**, 443–74.

1995

Edwards, A.W.F. Fiducial inference and the fundamental theorem of natural selection. XVIIIth Fisher Memorial Lecture, London, 20th October 1994. *Biometrics* **51**, 799–809.

1997

Frank, S.A. The Price equation, Fisher's fundamental theorem, kin selection, and causal analysis. *Evolution* **51**, 1712–29.

Lessard, S. Fisher's fundamental theorem of natural selection revisited. *Theor. Population Biol.* **52**, 119–36.

2000

Edwards, A.W.F. Sewall Wright's equation $\Delta q = (q(1-q)\partial w/\partial q)/2w$. *Theor. Population Biol.* (in press).

118

INDEX

additive genetic variance, 13
allele, 1
areal coordinates, 21-2, 55-6
asyntenic loci, 3
autosome, 2-3
average effect, 11-15, 20, 36-7
    lines of equal, 74-5
    variance of, 15
average effect of the gene substitution, 20
average excess, 11-15, 19-20

Baum and Eagon's Theorem, 47-9

Cannings' Theorem, 82-4
chromosome, 2-3
convergence
    diallelic locus, 10-11, 18
    multiallelic locus, 44
    multi-niche models, 91-2
    two diallelic loci: with selection, 104-5; without selection, 97-8
    unisexual selection, 88
    X-linked locus: with selection, 82-4, 86; without selection, 78
coupling, 94

diallelic locus, 1
differential region, 76
diploid, 3
dominance
    none in a multiplicative sense, 2; diallelic locus, 11, 23, 29, 32-3; multiallelic locus, 45
    none in an additive sense, 2; diallelic locus, 27, 32-3; multiallelic locus, 45-6; triallelic locus, 51
dominance deviations, 14
dominance variance, 12
dominant gene, 2, 27-8

envelope, 24-6

equilibrium
    diallelic locus, 10-11, 29-31
    multiallelic locus, 37-8
    multi-niche models, 89, 92
    two diallelic loci: with selection, 100-5; without selection, 97-8
    unisexual selection, 88-9
    X-linked locus: with selection, 79-80, 85; without selection, 78
Equivalence Theorem
    diallelic locus, 18-19
    multiallelic locus, 42-3
Ewens' Theorem, 102-3

Feller's Theorem, vii
    part I, 54-5
    part II, 56
fertilization, 3
Fisher's Theorem, 10-11, 30-1
Fundamental Theorem of Natural Selection, 48, 107

gamete, 2
gametes, random union of, 4-6
gametic array, 6
gametic selection, 6, 11
gene, 1
gene frequency, 4
gene ratio, 9
gene substitution, average effect of the, 20
gene-frequency space, 18-20, 68-71
genetic variance, 11
    additive, 13
    diallelic locus, 13-17
    multiallelic locus, 36-7, 39-40
genotype, 2
genotypic array, 5-6

Haldane's Theorem, 79
haploid, 3
Hardy–Weinberg equilibrium, 7, 22-3
Hardy–Weinberg Law, 7

119

INDEX

Hardy–Weinberg parabola, 22–3
Hardy–Weinberg proportions, 7
Hardy–Weinberg Theorem, 6–7
heterozygote, 2
homogeneous coordinates, 21–2, 50, 71
homologous region, 76
homothetic conics, 51–2
homozygote, 2

identical by descent, genes, 1
independent segregation, 4, 94–5, 99

Kimura's Maximum Principle, 46–7

least squares, 13–14, 36
linear model, 13–14, 36
linkage, *see* linked loci
linked loci, 4, 94–5
Li's Theorem, 38–9, 49
locus, 1–3

Mandel's Theorem, 43–4
map distance, 94
mating, random, 4–6
mean viability
 change in: diallelic locus, 15–18; multiallelic locus, 38–9
 diallelic locus, 9
 equilibrium: diallelic locus, 18–19; multiallelic locus, 38
 increasing function: diallelic locus, 17–18, 33–4; multiallelic locus, 40–2, 44, 47–9; multi-niche models, 90–2; two diallelic loci, 102–5
 multiallelic locus, 36
 multi-niche models, 90, 92
 not an increasing function: two diallelic loci, 100–2; viabilities different in the two sexes, 88; X-linked locus, 80
 surface of: dialellic locus, 31–3; multi-niche models, 91–2; triallelic locus, 50–74 *passim*
 triallelic locus, 50
 two diallelic loci, 99
Mendel's Laws
 First, 1–3
 Second, 4, 94–5
migration, 93
modal points, 75
Moran's Theorem, 101–2
multiallelic locus, 2

mutation, 3

phenotype, 2
population trajectories, 64–7
post-selection curve, 23–6, 31
potential function, 20
pre-selection curve, 23–5
projective geometry, 21–2

random mating, 4–6
random union of gametes, 4–6
recessive gene, 2, 7
recombination fraction, 4, 94–5
recurrence relations
 diallelic locus, 9, 12–13, 19
 multiallelic locus, 36
 multi-niche models, 89, 92
 two diallelic loci: with selection, 99, 104; without selection, 96
 X-linked locus: with selection, 79, 85; without selection, 77–8
reference triangle, 21–2, 55–6, 61–3
regression, 13–14, 37
repulsion, 94
Robbins' coefficient, 97
Robbins equilibrium, 97
Robbins' Theorem
 part I, 96
 part II, 98

Scheuer and Mandel's Theorem, 17, 32–4, 40
selective coefficient, 4
sex chromosomes, 2–3, 76–7
stability of the equilibrium
 diallelic locus, 10–11, 18, 31
 multiallelic locus, 42–4, 49
 multi-niche models, 91–2
 triallelic locus, 57–64
 two diallelic loci: with selection, 104–5; without selection, 97–8
 unisexual selection, 88–9
 X-linked locus: with selection, 80–2, 85–6; without selection, 78
syntenic loci, 4

trilinear coordinates, 21–2, 55

variance
 additive genetic, 13
 dominance, 12
 genetic, *see* genetic variance
variance of the average effects, 15, 37
viability, 4

120